PSYCHOANALYSIS
AND THE SCIENCES

André Haynal

PSYCHOANALYSIS AND THE SCIENCES

Epistemology — History

André Haynal

translated by
Elizabeth Holder

THE
UNIVERSITY
OF California
PRESS

BERKELEY • LOS ANGELES • NEW YORK • OXFORD

#26769102

Acknowledgements:

For excerpts: reprinted by permission of the publishers from *The Clinical Diary of Sandor Ferenczi*, edited by Judith Dupont. Cambridge, Mass.: Harvard University Press, Copyright © 1985 by Payot, Paris, by arrangement with Mark Paterson; © 1988 by Nicola Jackson for the English translation.

For excerpts: reprinted by permission of the publishers from *The Freud-Ferenczi Correspondence*, edited by Eva Brabant, Ernst Falzeder, and Patrizia Gianpieri-Deutsch, under the supervision of André Haynal. Cambridge, Mass.: Harvard University Press; by arrangement with Mark Paterson and Judith Dupont. To be published in 1993-95.

For excerpts from: Freud, Sigmund; *The Freud/Jung Letters. The Correspondence Between Sigmund Freud and Carl Gustav Jung* (W. McGuire, Ed.). Copyright © 1988 by Princeton University Press. Reprinted by permission of Princeton University Press.

The University of California Press
2120 Berkeley Way
Berkeley, California 94720

Copyright © 1993 by André Haynal

Library of Congress Cataloging-in-Publication Data

Haynal, André.
   [Psychanalyse et sciences face à face. English]
   Psychoanalysis and the sciences : epistemology—history / André Haynal.
      p.   cm.
   Includes bibliographical references and index.
   ISBN 0-520-08299-0
   1. Psychoanalysis—Philosophy.   2. Science—Methodology.
3. Science and psychology.   4. Freud, Sigmund, 1856-1939.
I. Title
   [DNLM: 1. Psychoanalysis.   2. Psychoanalytic Theory.   3. Science.
WM 460 H423p]
BF 175.H3848   1993
150.19'5'01—dc20
DNLM/DLC
for Library of Congress   92-48742
                                                              CIP

CONTENTS

v

. . . the *wide-angle* view

PART TWO
The historical dimension
in the construction of psychoanalytic theory

Photo: Hungarian figurine in Herend porcelain, nineteenth century, from the collection in the Musée de l'Ariana, Geneva (photo Jacques Henry, Vésenaz/Geneva)

PREFACE

Science and psychoanalysis regard each other like china dogs: tense, critical, almost hostile. Scientific ideals and psychoanalysis both exist. The aim of the book is to address this complex relationship by taking up two questions: how is the psychoanalytic "*knowledge*" that is presented in the literature actually *built up*? On what methodological and epistemological lines is its theory constructed, and what are the *historical* circumstances (the cultural and ideological history) that have made psychoanalysis what it is? These two fundamental questions—the relationship between the china dogs—are the theme of this book.

Its preparation aroused a very special sense of gratitude in me for all those who have accompanied me throughout my life and especially in my intellectual development. Any attempt to make an exhaustive list of them would obviously be doomed to failure. Only an autobiography could realize all the debts I owe.

In the first place I owe an enormous debt of gratitude to my scientific assistant, Maud Struchen, who took charge of the unenviable task of the typing and the bibliographical searches, as well as numerous other diverse jobs in the preparation not only of this manuscript but of all the previous ones. Marie-Christine Beck, psychologist and psychotherapist responsible for the Balint Archives in the Department of Psychiatry of the University of Geneva, gave me every possible help.

I would like to express my gratitude to the Ariana Museum in Geneva and to its Curator, Mme Marie-Therese Coullery, for having so willingly allowed me to reproduce the very beautiful dog on p. viii of this work, and also to M. Jaques Henry who photographed it.

I want to express my warmest thanks to Cesare Sacerdoti of Karnac Books and Klara Majthényi King of Communication Crafts for the wonderful job done in preparing the publication of this book, with particular attention to all the details. My thanks go to the University of Stanford for having twice invited me (1980–81 and 1988–89) as a "visiting professor". This provided an intellectual climate, both peaceful and stimulating, that favoured the formulation of my thinking. I particularly want to express my profound gratitude to the late Dr. Thomas A. Gonda, who, at the time of my first visit to Stanford, was Chairman of the Department of Psychiatry and behavioural Sciences at the Medical Centre of that University; to Dr. Bert Kopell, Acting Chairman, his successor; to Drs. W. Stewart and C. Barr Taylor, professors in the Behavioural Medicine pro-gramme, to whom I owe my second visiting professorship in 1988/89. This stay with my family remains an unforgettable experience, thanks to the friendship of Professors Annelise F. Korner and her husband, the late Sumner M. Kalman (both at the University of Stanford) and of Professor Wolfgang Lederer (Berkeley) and Mme Ala Lederer-Botwin (San Francisco).

Last but not least I would like to pay homage to my wife, Véronique, who endured the multiple preoccupations of her husband and always did her best to ensure a pleasant and creative lifestyle that helped considerably.

During the reflections that preceded the writing of this work, I realized that my occasional writings, prepared for various congresses or journals, are icebergs marking the visible points of a line of thinking that has been preoccupying me for many years. Ideas were formulated in these papers that it would have been difficult to reformulate in any other way. I have therefore made use of some of my earlier texts, not exactly in their original form, but elaborated within the framework of this book. For instance, my ideas on seduction in analysis were outlined at the Jerusalem Congress of the European Psychoanalytic Federation in 1977 (Haynal, 1977); those on trauma at the psychoanalytic Conference in Budapest in 1987, the first openly psychoanalytic international reunion in that country after forty years of silence (Haynal, 1989a). The formulation of my ideas on the "melting-pot" of psychoanalysis was stimulated by an invitation to a conference during the Duino Colloquium organized by the European Centre for Culture in 1983. The chapter on Ferenczi benefited greatly from an article developed with Ernst Falzeder on the relationship between Freud and Ferenczi [Falzeder & Haynal, 1989). The final chapter was the theme of the ceremonial "Sigmund Freud Lecture", which I was invited to give in 1979 at the University of Vienna, Austria, on the occasion of the fortieth anniversary of the death of the founder of psychoanalysis.

The questioning of experience has always been a feature of my professional development. My years of training in Zurich were in the "classical" current of psychoanalysis. This gave way to the challenge coming from the French psychoanalytic groups, both from the Paris Society, represented by Serge Lebovici amongst others, and from the French Association, through Didier Anzieu, Jean-Bertrand Pontalis, Victor Smirnoff, and Daniel Widlöcher. On the clinical side, the representatives of British psychoanalysis were the ones who exerted the greatest influence over me—viz. Michael Balint, Donald W. Winnicott, Wilfred Bion. I found the work of Mardi J. Horowitz, George Pollock, Robert J. Stoller, and Robert S. Wallerstein very stimulating. But it was Michael Balint who helped me to discover the importance of the historical dimension in our profession and the psychoanalytic traditions of my country of origin, especially the work of Sándor Ferenczi.

My interest in the evolution of the human sciences, which began in adolescence and which I have never relinquished since studying philosophy, was extended by a passion for linguistics and, more recently, for information theory, which posed intellectual challenges to which I had to respond. They strongly reinforced my conviction that psychoanalysis has to rethink its position in the field of science and learning at the end of this century. The ambition of the present book is to contribute to this.

The tradition in Geneva—my adopted country, which I regard as welcoming, tolerant, and pluralistic—is personified in Raymond de Saussure. After a personal analysis with Sigmund Freud, he became a founder member of the Paris Psychoanalytic Society; then during his peregrinations and pilgrimages, he spent some time at the Berlin Institute in analysis with Franz Alexander, followed by a long stay in New York during and after the Second World War. He returned to Geneva, his birth-place, towards the end of the 1950s. Along with Michel Gressot, another prominent figure in the psychoanalytic life of Geneva, he influenced me through his untiring interest in the work of Piaget and other thinkers, philosophers, and linguists who at that time were not yet called "cognitivists".

Marcelle Spira, in her amazingly creative personal vision of Kleinianism, opened the way for a "sensory-perceptive libertinism" in the work of interpretation—a way towards hitherto unexplored horizons.

Born *out of* reflection, this is a book *for* reflection.

AUTHOR'S NOTE

The two parts of the book are relatively independent. The reader who wishes to peruse the historical part first can do so without difficulty. It is at the end of the Overture and Epilogue that the interdependence of the whole becomes clear.

PSYCHOANALYSIS
AND THE SCIENCES

The psychoanalyst and the psychoanalytic encounter

> How bungled our reproductions are, how wretchedly we dissect the great art works of psychic nature!
>
> Letter from Freud to Jung, 30 June 1909
>
> It is well known that no means has been found of in any way introducing into the reproduction of an analysis the sense of conviction which results from the analysis itself. Exhaustive verbatim reports of the proceedings during the hours of analysis would certainly be of no help at all.
>
> Freud, 1918b, p. 13

One of the obstacles in the furthering of our psychoanalytic knowledge comes perhaps from the lack of opportunity to confront the experiences we live through in our psychoanalytic encounters. Obviously, this touches on a private and intimate experience that reveals, among other things, our personal way of reacting. The same problems were encountered in the course of the analytic training and were often attributed to examination anxiety or to their aspect of "initiation rites". Freud himself was running into them, before

1

finally deciding to publish his cases and to submit them to *public discussion* (1905e [1901], p. 7); but the move from a private discussion to a public one cannot be made without difficulties. Undoubtedly such a move is facilitated when the discussion takes place in restricted groups in which people know each other and where the sharing of implicit ideas and presuppositions makes for greater complicity and lessens the urge to criticize. Seminars and work groups are like families: they function on the basis of shared and unspoken hypotheses. Over and above this basic dimension, the difficulty inherent in clinical discussions is perhaps the source of a flight towards already developed and clinically distanced theories. In short, the problem arises from the fact that psychoanalysis is a personal practice, intimate and private, and there is a need to seek, as Freud did, a greater and more profound understanding of this encounter and all that it implies.

Already in 1955, however, Glover showed that practice is not, as might have been expected, totally a reflection of the theory, since it is far from being unified and cannot therefore be counted upon for a consensus. The divorce between metapsychology and clinical psychology deplored by Anna Freud (1969), among others, is reflected in the small number of descriptions of psychoanalyses in the current literature. There are many more opportunities for studying the generalities of metapsychological theory than the concrete unfolding of the psychoanalytic process, which involves us more. It is also easier to expound on the cases that seem paradigmatic to us than on those that are a cause for discomfort. However, the case of Dora, and others of Freud's cases, show clearly the interest that could be derived from discussing our difficulties and our failures instead of presenting psychoanalysis as a ready-found solution.

Describing or transcribing: the non-verbal in the session

The situation is paradoxical. In order to understand fully and to describe the session, the psychoanalytic encounter, it would be necessary to include the associative reverberations, in all their intimacy, of *both* protagonists. In other words, the "material" is not merely the *manifest* text; it also encompasses the associative links in the minds of both people present. But the revelation of this intimacy, the loss of protection of such a situation, risks bringing about a sort of "libidinal haemorrhage", a disengagement, or even exhibitionism. It is difficult to combine intimacy and revelation, bearing in mind all that can be hidden in the wish and the fear of confronting one's peers in the rivalry of different analytic styles.

This raises a fundamental question: some people would like to have as *objectifiable* an image of the psychoanalytic situation as is possible—even an audio-visual recording. At the cost of an enormous loss of information coming from associative and mnemic backgrounds, they hope in this way to have at their disposal material capable of provoking discussion about the theoretical basis of our clinical work or of the psychoanalytic conceptualization. A good deal of research, especially in the area of psychotherapy, has followed this approach. If what is important in the psychoanalytic encounter is all that is manifest *behind* the phenomena, in the form of *resonances* in the mental apparatus of each of the protagonists, such an approach may present merely an illusion. The *dialogue* between the two must be understood in the full sense of the term: a verbal and non-verbal dialogue, including those hidden forces—associations arising from the memory—that determine it.

This also brings me to the problem of the *writing* of psychoanalysis. Freud had chosen a particular, very literary style, and he professed astonishment that it "read like short stories" (1895d, p. 150). We are all aware of the distance between the notes he took on the "Rat Man" (1909d) and the version that he published. [Reading both the notes that Freud took during this analysis and the works in which he subsequently published them, it is clear that he used the accumulated material for a theoretical demonstration of an obsessional neurosis (as compared with an hysterical neurosis) and only exposed those elements that seem paradigmatic: the history of one of the dreams, the co-existence of love and hate, the homosexual connotations, etc.]

The *public* nature of data is one of the fundamental presuppositions of science. Now, psychoanalysis has either to decide that the exchange is so intimate that it must remain hidden from the public, being a concern only for the initiated, or to consider, as Freud did, that such facts must be accessible to science and should be in evidence. In that case it must struggle openly with the problem of how an intimate experience can be discussed publicly and what it can teach us about "human nature" in its affective and unconscious functioning, and it must seek the means of doing so quite freely.

The story of my tie
(and what ensues)

Hyacintha usually comes to her session at the end of the day and is therefore one of the last. On that particular evening, having to go out afterwards, I receive her freshly shaven, dressed in a black suit, and with a rather bright tie. As soon as she arrives, she says: "I don't like your tie. . . . I hate it." As I, too, had doubts about the tie, I think: "Ah, well! She is right, perhaps it is a bit too much. . . ." She associates: "You are not as you usually are. I have the impression that you are not there for me. You are dressed up for someone else. You are not my father when you are dressed up like that. He never dressed up except when he was going out." I think that at that point I said: "With your mother", and she continued: "Yes, besides it meant that he wasn't interested in me, that he was going with someone else. You, too, you're going out with someone other than me. It's not for me that you are dressed up. I realize that you have a life outside our sessions. . . ."

When she says that I am dressed up for someone else, I suddenly think: "After all, she cannot know whether it is for her or not." More elegant than usual and better shaven than normal at the end of a day—that is a fact. But it suits her to emphasize it. In the end it suits me, too: the psychoanalytic taboo means that in any case I have no right to seduce an analysand, and it "doesn't enter my head" (consciously) to dress up for her. . . . But if this is at a reality level, is it the only reality? When I hear myself saying: "How do you know whether it is not for you?", there is a long silence, then she says: "In fact, I have no idea." This kind of insistence, seduction, in the long run makes me feel ill at ease. It is a result of a whole period of very dull sessions, in which nothing happens, and we have talked over and over again endlessly about her homosexuality and her search for a mother figure during several years of analysis. I continue: "You noticed me as a man—a sexed man. It is hard to bear the sexual difference between us. You would rather we were the same." I realize that it is not the first time that she has noticed my external appearance. On another occasion, also a Friday session, I had been dressed in a dark suit, and at the time she had associated to it: "I don't like black", and then "You are ready to go out, so you are not going to be listening to me." Associations between black and abandonment had predominated at that time.

Actually, several ideas had arisen in my mind at the beginning of the session: first of all, I had wondered if I were right to dress up, what it implied with regard to the continuity of the sessions, which should be a backdrop, always the same. My tie was perhaps rather too loud in the lakeside fog of a gloomy autumn day in Geneva. Then some personal associations: in fact, my wife would not have liked that tie, she thinks, in any case, that I do not dress as I should, etc.

Finally, there was her association to her father—and my pleasure that we were at last emerging from a long period of erotic homosexual and maternal fixation, with the endless demands associated with it.

This was especially the case because in the preceding sessions we had just touched upon the topic of secrecy. The analysand was, in fact, going out very occasionally with a man who did not know that she was homosexual, and her parents ignored the fact that she had a boyfriend. I asked myself, *what is she hiding from me*, who is supposed to know everything? She tells me about her homosexual experiences, about her friend, her work, about all her alleged reality. And I suddenly understand that what she is not telling me about is precisely an aspect of her relationship with me, namely her wishes towards me. Speaking of her father in this context acts as a defence: if I am the father, I immediately accept the taboo and am confined within it, settled in a psychoanalytic convention that avoids the experience. Furthermore, she has said to me several times: "I am completely safe." When I question this, asking: "How do you know?", she replies: "You have to remain neutral, you have no need to play the role of man with me." Talking about her father is one thing; finding herself in a room alone with a man is quite another. Men are "grandfather" figures, in the end non-sexed, unsatisfying, not present, and unavailable. The repetition of "being with father" is one thing; being in my presence is something else.

After a long period when only the mother was present in the session, father being excluded from it, I introduced the idea of the sexual being, the third person, having been enclosed for a very long time in a "dual" relationship with her.

This interpretation brings not only a relaxation, but the account of a journey during the course of which she went to see a man with whom she had occasionally had sexual relations and whom she saw for a few days in a small town abroad. In the account of this journey several aspects emerge which she has not previously mentioned. Immediately after the journey—it took place about a year ago—she had talked about the fact that she felt imprisoned with this man and rather disappointed at not having been "made welcome enough". On that occasion she alluded to a dispute: after the second day she had wanted to move out into a hotel. This time, she recalls the intimacy and pleasure—in contrast to the previous account—of some moments that she spent with this man. Then she recalls a childhood memory. A memory of an uncle who had taken her to the amusement park in her native town and later, when she was a young woman, had accompanied her and her sister sometimes to fêtes and balls.

This woman, who was in the habit of holding intellectual political discussions with her father and who repeated this mode of relating with me, her robot analyst, suddenly, through a few memories of her lover and her uncle, changes her tone. I am no longer that intellectual with whom she can discuss things, her mother being "too stupid to understand", just as most women were to her—particularly the woman with whom she had what she described as "a profound relationship", which in the end was filled with complications and suffering. Am I imagining it, or is her handshake at the end of the session more relaxed, and is she a less brittle person on leaving the session than when she came tonight? In any case, the next time she says she is pleased to come—a communication that I think I am hearing for the first time in such a direct fashion. She embarks on the complications with her man friend, who, I may add, is a recent one. She says how difficult it is to be in a direct relationship with him, because he might ask too much of her. Even his attentions upset her. When, for example, he asked her to dine with him, she was at first unable to accept, but then rang him back half an hour later and invited him to her home. "My hypersensitive, touchy side . . .", she says. They had a discussion. Then she falls back into a kind of indifference, her speech suddenly becomes less intense. I point out that she is talking as though it is about someone else, and yet it is my impression that she really experienced this relationship very intensely. She replies: "That is true. I have been sulking. During the weekend, since the last session, I have come to understand that our relationship depends on the two of us, and not only on him. For the first time I felt involved and responsible." Then she goes on to talk about a male colleague at work, whom she described as a real friend, especially in terms of sharing.

She recalls a dream about me: "There was a round table, you and me. On your left, in the middle, was Monsieur Roger. There were some people, D and Z, who know you well and are real friends of yours, and they were saying nice things about your private life and your marriages. A lively, friendly conversation, and then it went blank."

The development of the associations around these people lead me too far afield (where?—these are the inevitable limitations of an account that, otherwise, is endless . . .). She reports that these people were, in the dream, very friendly and well disposed towards me. Then she recalls her holidays, those I spoke about before, and says that at the time she had wanted to show me her photograph album. "I'm beginning to get closer to you, and I am frightened I shall suffer." I pick up the elements of her dream—on the one hand marriage and divorce—and the last sentence, "then it went blank". I take up her wish for intimacy and her fear of my rejection. "It is exactly that atmosphere that I want to guard against." She associates to the death of Sadat, and the extent to which it upset her; he was a man of peace who had approached Begin, and there you are, he came to a sad end. "Even my father admired him." This led her to talk about her father again, about the difficulties she had in being in touch with him and discussing things with him, and then his disappearance. I interpret: "Perhaps a growing closeness will bring about another disappearance—maybe my death."

Stones can talk

The account of this sequence shows, if demonstration is needed, that psychoanalysis is, in some respects, truly interminable—a fact that Freud had already noticed in a letter to Fliess as early as 16 April 1900 (1950a), many years before he wrote "Analysis Terminable and Interminable" (1937c). In fact, to achieve a more accurate understanding of these sessions, I should have spoken of our common past, of all that we have experienced together during treatment, of all that we said and didn't say and all that ensued from it. What is more, I would have to recount not only her life, *but mine as well* . . . Is this not the real reason that Freud found it ultimately impossible to talk satisfactorily about a psychoanalysis—and about affective non-verbal communication as well!? (Cf. the quotation at the head of this chapter.)

The psychoanalytic process is "the remembrance of things past", a search for and construction of memories. Such a remembering, of course, means finding "the words to say it", for the unconscious is not a language, it is a totality of sensations and imagined experiences, of images linked to moments of anxiety or desire. It is a matter, therefore, through language, of re-establishing these elements in a continuity and of giving them meaning in relation to the personal history of the individual. The *meaning* is linked to continuity—to coherence: restoring his history to an individual is a "narcissistic gain" for him, the re-establishment of his image of himself and of the meaning that is linked to it.

The dialogue and interaction of psychoanalytic encounters are linked to *the history* of the individual. Freud's allusions to *archaeology* (Freud, 1905e [1901], p. 12; 1937d, p. 259) have been little used for the understanding of psychoanalytic processes, yet it is clear that Freud was fascinated by this model. [Etymologically, the word "archaeology" has a double reference: that of the scientific exploration of the past, in its resemblance to the word "archaic", and to the suffix "arch", corresponding to the German "*Ur*", meaning the principal, or the main event or experience of the subject. The archaeological point of view, therefore, approaches not only the past, but also the primordial, the events determining a future outcome.] The archaeological–historical metaphor was present throughout his life. [According to Guttman's index (Guttman, Jones, & Parrish, 1980), the word "history" is found 569 times in the English version of his writings.] In his work, "The Aetiology of Hysteria", Freud (1896c) writes:

> If [the analyst's] work is crowned with success, the discoveries are self-explanatory: the ruined walls are part of the ramparts of the palace or a treasure house; the fragments of columns can be filled out into a temple; the numerous inscriptions, which, by good luck, may be bilingual, reveal an alphabet and a language, and, when they have been deciphered and translated, yield undreamed-of information about the events of the remote past, to commemorate which the monuments were built. "*Saxa loquuntur!*" [stones speak]. [p. 192]

And three years later (1899a), he writes:

> It may indeed be questioned whether we have any memories at all *from* our childhood: memories *relating to* our childhood may be all we possess. Our childhood memories show us our earliest years not as they were, but as they appeared at the later periods when the memories were aroused. In these periods of arousal, the childhood memories did not, as people are accustomed to say, *emerge*: they were *formed* at that time. [p. 322]

[Freud's clinical observations led to conclusions that correspond totally to what contemporary science takes as established, after partly experimental, careful and precise researches (see later).]

Much later (1916–17), he wrote:

> If they [these events] have occurred in reality, so much to
> the good; but if they have been withheld by reality, they are
> put together from hints and supplemented by phantasy.
> The outcome is the same, and up to the present we have
> not succeeded in pointing to any difference in the conse-
> quences, whether phantasy or reality has had the greater
> share in these events of childhood. [p. 370]

He returns to this idea throughout his work, up to his paper on
"Constructions in Psychoanalysis" (1937d):

> Instead of that, if the analysis is carried out correctly, we
> produce in him an assured conviction of the truth of the
> construction which achieves the same therapeutic result as
> a recaptured memory. [pp. 265–266]

He defines the construction as putting before the patient's eyes
a piece of his early history that has been forgotten (p. 257). A
little further on he stresses:

> Just as our construction is only effective because it recov-
> ers a *fragment* of lost experience. [p. 268, italics added]

Archaeology is not yet history. It differs from it in that it is
supported by objects, not constructed on a discourse, and that
these objects are only points of departure or of emphasis in a
discourse yet to be spoken. As the account of the subject's
history emerges, it is reorganized around the main themes of
wishes with the conflicts arising from them—pre-history, therefore, belonging to a time *prior* to the organization of the discourse, before the organization of historical documents or a
structured discourse. The attainable past belongs to the preconscious, the reconstructible past to the unconscious.

History thus becomes a fictitious, imaginative, creative, and
more coherent account than the materials on which it is based.
During this encounter, the individual's history is organized
through the psychoanalytic encounter; he takes upon himself
the *responsibility* for his history, identifying himself with a history that he makes his own. By recovering the past that he had
rejected, he, in a way, composes a sort of new history, a new
past. What is lived through on the one hand, what is worked

through on the other, follow a development that is uninterrupted and never-ending and bears the stamp of the individual.

Working-through does not always follow experience, but it
sometimes precedes it. Anzieu (1975) shows to what extent
many of Freud's concepts were already present at the time of
his self-analysis, although of course there is no denying the
importance of the radical revisions that occurred later.

Interpretation does not exhaust the content. It is also a
strategic action that, in the course of a developmental process,
is determined by the progressive and sometimes regressive
needs of that process. In fact, the pluralism of psychoanalytic
styles shows that, during its growth, psychoanalysis has encountered different personalities, cultures, and a variety of
sensitivities, developing widely diverse styles that are reflected
in the different conceptualizations. However, certain principles
underlying every psychoanalysis, whose existence is not the
exclusive prerogative of any one school, enable us to determine
whether or not a process is a psychoanalytic one.

Even if it is conceded that at some moment theory precedes
practice, it does not determine it rigidly, and practice must
remain the creative working-through of an experience. *Confin-
ing* practice within a theory risks bringing about the exclusion
from it of any element of surprise or possible openings towards
the emergence of new connections and insights. Some specific
ideas, for example that of *"Takt"*, come to mind in this context
("Takt", Freud, 1926e, p. 220; Ferenczi, 1928 [283], pp. 89,
100). Freud's German word *"Takt"* refers to the pleasant touch
and "rhythm" of a musician's performance, in which aesthetic
considerations are more important than the rational qualities.
Since a musical score can be rendered in very different ways
according to the player or conductor, so the psychoanalytic
session will be experienced and therefore expressed—interpreted—differently according to the *analyst's sensitivity*. The
analysand's discourse is not only a text, but also a communication by means of non-verbal signals: tone, rhythm, gesture,
melody, pauses, sighs, laughs, and there are unorganized
texts, a slip of the tongue, a dream fragment. All these signs
become clear only when the word has apprehended and explained them in a symbolic dialogue, permitting their integration in the conscious experience of the subject.

The interpreter is not satisfied with merely translating a language, of rendering it *comprehensible* to those for whom, previously, it was not. Similarly, an actor interpreting a role, a conductor conducting a symphony, do more than simply render a text; they translate it with all the nuances of *how* they heard what they heard. The situation and setting of the psychoanalytic encounter permit the creation of the necessary space *for* this interpretation, which is the analyst's contribution to the analytic process. Its central importance stems from the fact that it gives his functioning a specificity in the psychoanalytic encounter. In this perspective, neutrality (non-implication—refusal to take part—in a conflict: the expression probably comes from the Zurich school, with reference to Swiss neutrality; in the original edition of Freud's works, this expression is *nowhere* to be found) is one of the preconditions for the creation of the analytic space (Viderman, 1970), of the interpretation that maintains that space, and, as a result, of the process that unfolds within that space.

If the aim of interpretation is to revive the subject's associative processes in the analysis by giving a meaning to his experience, it also facilitates the *partial disengagement* and setting up of an *optimal tension* for the protagonists. From this viewpoint, the process is therefore a creation of two partners and interpretation is an instrument fundamental to the diminution of internal obstacles (interpretation of resistances). It also enables moments of relative relaxation to be found in order to pursue that experience, that voyage, which is an exploration of the analysand's inner life through his encounter with the analyst. It should not be forgotten that Freud himself was very sceptical about the possibilities of self-analysis (Freud, 1912e, p. 116; see also "Editor's remarks", 1915e, p. 202); the dimension of the psychoanalytic encounter is therefore vital with regard to the implementation of the analysis.

The psychoanalytic dialogue permits rediscovery of the past. It is not only an account of the past, it is a gradual *recreation* of it, as that past is articulated in the interaction between the analyst and the analysand—a recreation under the modifying influence of the analyst to whom the discourse is addressed. The analyst therefore contributes to the creation of the past. The analysand *and* his analyst interpret the past: the important elements that are relived are experiences, interactions, encounters—real or imaginary—rediscovered in the form of dialogue. The transference is heir to them, *heir to a former dialogue* understood, reconstituted, and constituted in the psychoanalytic dialogue.

Interpretation is the discovery of the meaning and intentions of an internal, subjective, very real world: a royal road towards the past, towards the world of the imagination and desires.

But the psychoanalyst is also the *container* of the analysand's *drives*. If it were only a matter of discourse and its interpretation, his task would not be so difficult. But the psychoanalyst rapidly becomes a container of drives—*father or mother in the transference*: *who am I* for the patient? He is at the focal point of passionate forces—that is, the specific aspect of analytic attention. Freud (1915a, p. 168) emphasizes that transference love is "genuine love". The language in the psychoanalytic session conveys instinctual experiences. Psychoanalytic discourse is distinguished from other dialogues by the fact that it plumbs the depths of instinctual experiences. The psychoanalytic situation is not a place where experiences are analysed from without; it becomes the *place of experiences*. Transference is not "like love"; it *is* love, fixed and immutable. It is here that the difficulty of *mourning* can take place, and the resistance to changes (often linked to early trauma or to early losses; these are the people it is convenient to label "narcissistic personalities", because their lack of trust in object-related experiences conditions their withdrawal to a narcissistic level, to preoccupation with the self).

As Freud liked to emphasize, it is not through its lack of authenticity that the psychoanalytic situation differs from external life, but through its *proposed aim*—namely, to understand in the present the resurgent elements of former imagos (Freud, 1914g). It is a delicate balance, since this must not constitute a resistance against "living"—either on the analyst's part or on the analysand's part—but must form a dialectic ebb and flow between "living" and "understanding". Sterba's "therapeutic disassociation of the ego" (1934) and Reik's "third ear" (1948) are metaphors that try to describe this dialectic, a dialectic of talking and listening, of being both actor and spectator, in something "genuine", in the most profound experience, while at the same time seeking to decode the one or, rather, the various possible meanings in relation to the individual's particular history.

It has often been said that interpretation is also a means by which the analyst can distance himself from the material with which he is presented (Szasz, 1963; Flournoy, 1968), it is in some way his defence for maintaining the psychoanalytic process (Freud, 1925d [1924], p. 27). ". . . As she woke up on one occasion, [my patient] threw her arms around my neck. . . . I was modest enough not to attribute the event to my own irresistible personal attraction . . ."—in this way began the history of psychoanalysis.

Freud regards the transference as *a false connection*, a false relationship (1895d, p. 303). The countertransference means involvement in the psychoanalytic dialogue; Freud grasped the importance of it as early as 1912. The analyst must, he said, make of his unconscious a telephone receiver for the patient's unconscious that emerges (Freud, 1912e, pp. 115–116). [In order to be a "receiver", the analyst must, according to Freud, *repress* the disturbing elements of his countertransference: however, he seems to reserve the notion of countertransference to the negative elements; when he speaks of its positive aspects, he does not use this term—for example, in the image of the telephone receiver.]

The psychoanalytic situation lies between the two great founding myths of Narcissus and Oedipus; or, to paraphrase Balint (1968, pp. 24–25, 43), it is a to-and-fro between the two people present: a "one-person" situation with the denial of the other's presence; a "two-person" situation feigning the elimination of the absent third one, to whom, however, the analysand addresses himself; and, finally, a three-person situation. The different conceptions of technique, which turn around the problem of narcissism, are linked to the two-person relationship with reference to the triangular one—a relationship with the mother, concerned with the introduction of the father—whether the two protagonists are seen as a constructive mirror (Kohut, 1971) or as a death-bringing one (as in the Narcissus myth), or as 3 *minus* 1, with the absence of the third. That latter presupposes that the interpretative grid is centred on the

Oedipus complex, as Freud anticipated in one of his inspired intuitions (1905d).

The analyst is sometimes a narcissistic double of the analysand, and sometimes the object of his wishes. "You would like me to feel what you feel", expresses the demand for total understanding, for symbiosis, the mirror image. "You would like to have from me what you wanted from your father" (or "from your mother") signifies the demand for gratification of a wish in the object relationship. In order to formulate and understand what is going on, it is necessary to grasp the "theme" of the text of the image underlying the text. What are the analysand's wishes (drives), what does he fear in relation to these drives (signal, anxiety), and what defences will he set up?

If psychoanalysis is not turned into an exercise of omnipotent therapy, an entrenchment behind benevolent neutrality, it can become the encounter between two human beings, two individuals who will try to express themselves and even to care for each other (Searles, 1975)—so much so that analysts can well be said to have as much need of their analysands as the analysands have of them.

This encounter could also be the basis of the investigation of the epistemological issue, i.e. of searching for a way of constituting out of these experiences a theory, either in the perspective of proximity (theory of the encounter itself) or at different, more distant levels.

PART ONE

. . . THE *CLOSE-UP VIEW*

CHAPTER ONE

The psychoanalytic process:
a symphony in three movements

Scientia vero, quae aliis traditur, eadem Methodo (si fieri possit) . . . est insinuanda, qua primitus inventa est.
[Knowledge, for its transmission to others, should be as nearly as possible in the form in which it was discovered.]

Francis Bacon, 1620

My paradigm of the psychoanalytic process, like any other paradigm, is a limited one. Its aim is to clarify the *major* points of impact in treatment.

It highlights three "*moments*" of emotional impact in psychoanalytic treatment, which, according to their specific *impact*, risk being treated as taboos, repressed, or split off:

1. the moment of mutual *seduction* (when sexuality emerges with all its concomitant guilt);
2. the moment when the *trauma* reappears or, in more general terms, when the subject is re-exposed to the fundamental *difficulties* in his life (i.e. everything with a painful, anxious,

19

or depressive connotation through its link to failure, to what has fallen short of the ideal, which is therefore likely to be pushed away, not grasped, expelled from communication);

3. the process of "working through", of transformation, of change (these difficulties culminate, as it were, in trauma and its consequences), which refers to loss and mourning to be done—and, ultimately, to the symbolism of Death.

Seduction

If seduction is defined as an active movement of solicitation, of instituting contact and the growth of *intimacy*, it is the definition of an affective phenomenon that is communicated through preponderantly emotional, non-verbal channels (looks, gestures, voice, posture, etc.). The setting-up of an analytic bond, and thus the institution of the *process* of every analysis, takes place in a *founder*-movement of mutual seduction. [I expressed this idea already in 1983, at the European Psychoanalytic Federation meeting in Jerusalem (Haynal, 1983a). Later, in his book, *New Foundations for Psychoanalysis: Primal Seduction*, Jean Laplanche (1987) wrote along similar lines, although his level of methodology and conceptualization are different from mine.] The analyst offers a presence, an intense listening, an intimacy—the analysand his presence, his honesty, his expectations, and his demands. In this *perspective*—of Bion's "vertex" (1965, pp. 106–107)—the affective–emotional setting-up of the process can be grasped. Freud's preoccupations in his earlier works are of the same order: seduction and sexuality are the first themes mobilized by the encounters with his analysands, themes from which he will part, in that adventure he calls psychoanalytic treatment.

Freud's reference to "the therapeutic alliance" (1937c, p. 235) emphasizes other aspects closer to the rationality of the "system ego". According to post-1923 Freud (1926d [1925], p. 154), the analyst offers "assistance"; he is the ego's ally (Freud, 1940a [1938]):

The Ego is weakened be the internal conflict and we must go to its help. [p. 173]

The analytic physician and the patient's weakened Ego, basing themselves on the real external world, have to band themselves together into a party against the enemies, the instinctual demands of the Id and the conscientious demands of the Super-Ego. . . . We form a pact with each other. . . . This *pact* constitutes the analytic situation. [pp. 173–174, italics added]

The method by which we strengthen the weakened Ego has as a starting-point an extending of its self-knowledge. . . . The first part of the help we have to offer is intellectual work on our side and encouragement to the patient to collaborate in it. [p. 177, italics added]

On the patient's side, a few rational factors work in our favour, such as the need for recovery . . . and the intellectual interest . . . in the theories and revelations of psychoanalysis; but of far greater force is the positive transference with which he meets us. [p. 181]

As is well known, the analytic situation *consists in our allying ourselves with the ego of the person under treatment*, in order to subdue parts of his Id which are uncontrolled—that is to say to include them in the synthesis of the ego. The fact that a co-operation of this kind habitually fails in the case of psychotics affords us a first solid footing for our judgement. The ego, if we are to be able to make such a pact with it, must be a normal one. But a normal Ego of this sort is, like normality in general, an ideal fiction. [Freud, 1937c, p. 235, italics added]

The seductive element permits an understanding of the emotions released by the following: the "honeymoon" (Grunberger, 1971), in the sense of the elation linked to the (partially, incompletely) "successful" seduction, of the awareness of the wishes and hopes, and equally the fears, anxieties, and profound concern aroused in both protagonists.

The child, in its helplessness (Freud, 1950a [1895], p. 318), resorts to seduction and tries to activate the people in its environment to fulfil needs and desires. The new research on babies has amply demonstrated the extent to which they are active in soliciting the help that they need (Stern, 1985); this is done affectively and is in the nature of "seduction". We tend to use the same strategy (often unconsciously) when we find ourselves in situations where there is a feeling of deprivation.

The psychoanalyst creates an *intimacy*. In an interaction initiated by the psychoanalyst, the analysand will obviously contribute to the creation of that intimacy. This is the first step, with its consequences: the setting-up of the ego ideal (the appearance of extraordinarily high, even inordinately megalomanic hopes), and other so-called "regressive" manifestations, in particular some emotional dependence (which has always been the target of critics of psychoanalysis—the latest version of this is Masson's, in 1988, in which he talks about "emotional tyranny").

It is quite clear that focusing our thinking and elaboration on the *emotional encounter* and what it mobilizes will lead to a conception of psychoanalysis in which the *experiencing* of that emotionality and its ultimate, if imperfect, analysis will play an important role. The clarification of, and emphasis on, this aspect of psychoanalysis was historically the work of Sándor Ferenczi. Freud put more stress on "insight", on "analysis" in the strict sense of the term, in other words on what today is called the "cognitive" aspect. In his wish to bring emotionality into that same "insight", Ferenczi thought that *experience* must precede it with as great a depth as possible. Opposing "insight" [*"Einsicht"*] and "experience" [*"Erlebnis"*] seems to me grossly simplistic. It can only be a question of *degrees* of experience; for example, the question discussed by Freud and Ferenczi (1990, Fer. to Fr., 5 December 1931; Fr. to Fer., 13 December 1931; Fer. to Fr., 27 December 1931—the as yet unpublished correspondence between Freud and Ferenczi uses "Fr." for Freud's letters and "Fer." for Ferenczi's), of knowing what is tolerable in the manifestation of sexual elements (and their inevitable discharge in some form), and to what extent that *discharge*—such as the "little kiss" of Ferenczi's patient which initiated an important controversy [today we know that without such a phase, without that "denunciation" of Ferenczi to Freud by the analysand, an important aspect of the case could perhaps not have been analysed: the father's denunciation is an infantile theme in that patient's history of which she was never aware (Ferenczi, 1985 [1932], p. 11)]—or emotions generally, should be re-experienced only *minimally*, and for the greater part frustrated by the framed work and by the very strict non-responsive attitude of the analyst, which was Freud's position. [Cf. the adage according to which the analysand can "say anything and do nothing" in the analytic situation (Fr. to Fer., 13 December 1931).] The historical perspective of today shows clearly that this is a question of *degree*, and it might be supposed that the antagonism projected by the analytic world in this dialogue between Freud and Ferenczi brought about the adoption of an impassioned support for Freud and the ostracism of Ferenczi as a "dissident" (I shall return to the meaning of this expression, which facilitated the elimination of the problem by creating a "memory gap" in history, touching, as it did, upon the analyst's emotions and therefore upon a vital personal dimension), positions linked to the difficulty of the problem but which obviously did nothing to resolve it. Historically, ego psychology has retained the idea of the "*a minima*" as it appears in the formulae of the ageing Freud. [This was the period when Freud had more pupils than patients, when he claimed to be unenthusiastic as a therapist, etc. (on this issue, see Haynal, 1987a, pp. 3, 10, 32–33, 140).] The expression of *emotionality*, the importance of experience, and the idea of an analytic framework that permits the maintenance of such experience and regression find continuity in the heritage of Ferenczi among all those who have worked in the *object-relations* perspective: the British "middle group" (Winnicott, Balint, Rycroft, and others), the Kleinian group, and various other groups (Searles and others) in America, France, and elsewhere. There can be no doubt at all that the majority of psychoanalysts are heirs to this conceptualization—here, too, to varying degrees.

As a point of departure, seduction enables us to confront all aspects of *emotionality* from the beginning of treatment, which will subsequently flare up in the course of the analytic process.

The problem of the *event* has not often been addressed as such in psychoanalysis. The primary importance that Freud accorded to the *past*, in particular to the *psycho-sexual* past, is largely responsible for it. Much later on, he examined one single event, the loss of the libidinal object (Freud, 1915b, 1916a, 1917e [1915]). Interest in the *present* arose along with the development of technique; clarification of the present in terms of the past plays an increasingly important part. Research on "life events" (Brim & Ryff, 1970; Dohrenwend & Dohrenwend, 1974; Holmes & Rahe, 1967; Lindenmann, 1944) has, in turn, shown how those events have influenced the psychic development and state of health or illness of the individual. The notion of "stress", from a psychoanalytic viewpoint, becomes one of traumatizing elements, probably because of the force they draw from the past, in primitive traumata. In fact, the Freudian and Ferenczian theory of trauma may well further our understanding.

Trauma

A psychoanalytic fantasy, often expressed in private, is of meeting the perfect analysand: attractive and full of charm, able to express himself with ease, one who finally proves to be a subject without any major problem. . . . Schofield's formula of the YAVIS-syndrome (1964) exactly describes this wish: "Young, Attractive, Verbally fluent, Intelligent, Successful." But the reality is quite different, and not without *good reason*. In every situation in which the subject *hopes* to be able to resolve his difficulties, he cannot economize on the re-experiencing of his problems, of the traumatic forces that hold him in check. Psychoanalysis *exposes* the subject to these aspects just as they are inscribed in his structure, in his psychic life. This leads us to consider the role of traumatization in the psychoanalytic treatment as a straightforward process occurring after the removal of initial resistances. To be able to talk about those processes and re-live them emotionally is a relief—experimental psychology itself demonstrates it (Pennybacker, 1988). If *seduction* created the link that makes togetherness possible after the analyst's offer and the analysand's demand, it is the trauma that will lead us to the centre of the process, to the *repetition* of painful, sometimes even tragic, aspects in the subject's life, the analysis being a digest of different, difficult, even traumatic earlier experiences with primary external "objects".

I have emphasized the first affective contact and attempted to understand it through the notion of seduction, and its implication for the development of treatment—viz. the process and its motivations. It is clear, however, that we must be aware of the *difficulties* the subject will meet in the course of that same process. The painful, embarrassing, shameful, and guilty nature of what *arises* (all that can conveniently be called "associations") clearly shows that it is the emergence in the "*here and now*" of distressing moments from the past, of the subjects most profound problems. It is a *re-exposure* in the *relationship* to the most painful problems in his past by virtue of what psychoanalytically is termed the "repetition compulsion". In order to explain this mysterious force, Freud invoked the "death instinct" (Freud, 1920g). It is not my intention to discuss this here; it would require too great a digression, even if the viewpoint of some contemporary biologists, such as Henri Atlan (1979), opens up some interesting considerations on the basis of information theory. I am merely emphasizing the fact that psychoanalysis aims at "*breaking through*" the destructive quality of the repetition by enabling the subject to realize his desires, sometimes in a sublimated way if the primary aims cannot be attained because of reasons linked to assumed values embodied in the superego or ego ideal. Or, on other occasions, it facilitates the "dissolution" of underlying conflicts, though rather less often than is optimistically supposed in overestimating the forces at work in psychoanalysis. (Freud uses three different words for "destruction": "*vernichtet*", "*zugrunde geht*", and "*zerstört*"; Strachey translates these with the single word "destroyed". Various different terms are used in current French translations; however, they all, without a doubt, refer to the same underlying, more or less explicit, concept.) In any case, it helps the subject towards a better mastery of those elements that lead to repetition, and therefore of repetition itself. I shall first address this topic as it has been formulated in psychoanalytic theory under the headings of trauma, its repetition in treatment, and its outcome.

The *emotional experience* of trauma plays a fundamental part in psychoanalytic treatment. Recognition of this is not new. Already in 1896, Freud emphasized that "symptoms . . . can only be understood if they are traced back to experiences which have 'traumatic effect'" and "refer to the patient's sexual life . . . in early childhood (before puberty)" (1896b, p. 163). The Greek word "*trauma*" means a wound ["where the skin is broken", as Laplanche and Pontalis rather pedantically define it in *The Language of Psychoanalysis* (1967)] and presents a perspective in which we, the neurotic ones, are all injured, because of wounds in an area where our deepest, most visceral desires are rooted, desires in ceaseless continuity from the beginning of our existence. Later traumas arouse mnemic traces of earlier infantile ones. These latter may remain unconscious through repression after the freezing of affect (usually anxiety). [The trauma is always present, infiltrating the personality, as Freud envisaged it, under pressure to take possession of consciousness. Robert Stoller (1975, 1979) has shown the extent to which even the basic sexual fantasies of the individual penetrate it and are constantly present to be improved and put right.]

Freud's next step, in 1897, is well known: the change of perspective expressed in the famous phrase: "I no longer believe in my neurotica" (letter to Fliess, 21 September 1897, *in*: Masson, 1985, p. 264). Having realized that the scenes of *seduction* were often invented by his patients, he introduced the idea that *traumas* were linked to *fantasies* of desire and that it is a matter of indifference knowing "whether these elements correspond to reality. . . . The result is the same" (1916–17, p. 370). The trauma, then, is situated at the level of an emerging *wish*, the satisfaction of which clashes with "reality"—that is, with the reality of the Other and of others. Is this formulation a betrayal, as Masson claimed in 1984? The sentence quoted follows Freud's warning against the danger of under-estimating the importance of the *actual* seduction. In this context it is worth emphasizing that Freud's position does not imply the *negation of reality* of some traumas, while acknowledging that others are formed from "hints" ["*Andeutungen*"] completed by "fantasy" ["*durch die Phantasie ergänzt*"] (1916–17). He extends the field covered by this concept and introduces the idea of the wish and the fantasy of the wish, without casting doubt on the reality of gross sexual traumas.

The ego feels helpless ("*hilflos*", Freud, 1926d [1925]), abandoned to the situation, to a flood of *excessive* excitations, too powerful for the mental apparatus to deal with. [In this sense, the notion of trauma is essentially an "*economic*" concept (Freud, 1916–17, p. 275). It can never be sufficiently stressed that "trauma" and the re-living of it do not represent only one moment of transitory anxiety, but also a profound experience of helplessness or dereliction (Haynal, 1987b).] In another formulation, Freud asserts that "A person only falls ill of a neurosis if his ego has lost the capacity to allocate his libido in some way" ["*die Libido unterbringen*"] (Freud, 1916–17, p. 386), which comes back to "helplessness".

As with other psychoanalytic concepts, that of trauma has gradually been enlarged. [In the history of psychoanalysis, one often encounters this imperceptible and gradual widening of concepts. First of all a phenomenon is noticed and its importance recognized in the gross caricature of it as presented by pathology. Once attention has been drawn to it, it is found, *a minima*, in a whole number of analyses, even in all of them. The impact of the first experience may well be diminished, but this attests to the fact that the events linked to the construction of the mind are the same. In order to take into account the problem of differences, Freud resorted to the quantitative notion. This, however, remained speculative, since the quantification is not based on any measure but on subjective differences: "force", "energy" proved to be metaphorical expressions. Epistemology has not solved the problems raised by these difficulties, probably for the most part because the hermeneutic character of psychoanalysis was not recognized for a long time.]

The gross traumas of the period in which *Studies on Hysteria* (1895d) and "Further Remarks on the Neuro-Psychoses of Defence" (1896b) were written gave way later on to more hidden and sometimes more insidious traumas. In the network of events, wishes, and fantasies involved in this concept, the emphasis is on one or the other according to the sensitivity of the analyst, and to the life and personal history of the analysands. In addition to sexual trauma in the strict sense of the word, there are others such as the "abandonment" trauma of Germaine Guex, René Spitz's hospitalization trauma, Bowlby's early separations, and others, which all seem to depend on clear-cut economic disruptions; they are serious because of the loss of love (other sometimes lesser traumas generate more subtle intrapsychic phenomena).

Trauma, then,

1. is linked to *wishes* that are frustrated. Whether or not the outcome will be a trauma is determined at the point where an event encounters the subjective images of a person's wish ("the state of his ego": the degree of his helplessness); beyond a certain level of trauma, nothing can resist it without suffering the consequences of it;

2. is clearly situated in the relationship to the *Other* (to the "drive object" according to traditional psychoanalytic terminology).

What is the reality of the Other? What is its impact on the psychic life of the subject? What capacities does the subject possess to *get through* a trauma?

For Ferenczi, it is the non-comprehension of adults, the "non-fitting", the non-correspondence that is traumatic: the moments when the child cannot feel understood. For the child seeking tenderness, the adult's sexual aggression and his passions create a paradigmatic, acute situation of *non-correspondence* of two wishes, the child's and the adult's. And that is only the beginning: subsequently the child will feel split between what he has understood and what is being said to him—between guilt and innocence, simultaneous complicity and resistance. "Almost always the perpetrator behaves as though nothing had happened" (Ferenczi, 1933 [294], pp. 162–163).

Ferenczi strongly emphasized the importance of sexual trauma: "I obtained above all new corroborative evidence for my supposition that the trauma, especially the sexual trauma, as the pathogenic factor cannot be valued highly enough"; and he adds, not without irony:

> even children of very respectable, sincerely puritanical families, fall victim to real violence or rape much more often than one had dared to suppose. Either it is the parents who try to find a substitute gratification in this pathological way for their frustration, or it is people thought to be trustworthy such as relatives (uncles, aunts, grandparents), governesses or servants, who misuse the ignorance and the innocence of the child. The immediate explanation—that these are only sexual fantasies of the child, a kind of hys-

terical lying—is unfortunately made invalid by the number of such confessions, e.g. of assaults upon children, committed by patients actually in analysis. [Ferenczi, 1933 [294], p. 161]

It is mainly the lack of communication *after* the event that renders it really traumatic, and it is *verbalization* that makes it possible to overcome this "flooding of excitation" that is characteristic of trauma.

As Balint (1969) recalls, Ferenczi describes the child as being, at first, full of confidence. A second phase follows when the adult offers "a highly exciting, frightening or painful experience" (p. 432), an intensive interaction, but one that, Balint adds, is not "in itself traumatic". However, in a third phase, the child wants either to continue the game or to understand and then comes up against a totally unexpected "refusal": the adult behaves as if he knew nothing at all about the earlier excitement.

With this idea in mind, Ferenczi's technical aim is to *allow* the subject to re-live the trauma (to re-experience it) and then, through *verbalization*, to achieve recall. It is the *experience* of *repetition* in the transference that is one of the fundamental elements of psychoanalytic treatment, making *understanding* possible. Freud recognized the importance of these ideas. For example, he wrote:

My attitude to the two books concerned, then, is the following. I value the joint book as a corrective of my view of the role of repetition or acting out in analysis. I used to be apprehensive of it, and regarded these incidents, or experiences as you now call them, as undesirable failures. Rank and Ferenczi now draw attention to the inevitability of these experiences and the possibility of taking useful advantage of them. [Freud, 1965a, 15 February 1924, p. 345]

Ferenczi (1933 [294]), however, already foresees various consequences of trauma:

Not only emotionally, but also intellectually, can the trauma bring to maturity a part of the person. I wish to remind you of the typical "dream of the wise baby" described by me several years ago, in which a newly born child or an infant begins to talk, in fact teaches wisdom to the entire family.

The fear of the uninhibited, almost mad adult, changes the child, so to speak, into a psychiatrist, and, in order to become one and to defend himself against dangers coming from people without self-control, he must know how to identify himself completely with them. [p. 165].

With regard to parents, he also notes that:

In addition to passionate love and passionate punishment there is a third method of helplessly binding the child to an adult. This is the *terrorism of suffering*. Children have the compulsion to put to rights all disorder in the family, to burden, so to speak, their own tender shoulders with the load of all the others; of course, this is not only out of pure altruism, but is in order to be able to enjoy again the lost rest and the care and attention accompanying it. [pp. 165–166; italics in original]

In order to explain the consequence of such occurrences, he introduces the concept of "splitting" (Ferenczi, 1930 [291]):

In every case of neurotic amnesia, and possibly also in the ordinary childhood amnesia, it seems likely that a *psychotic splitting-off* of a part of the personality occurs under the influence of shock. The dissociated part, however, lives on hidden, ceaselessly endeavouring to make itself felt, without finding any outlet except in neurotic symptoms. [p. 121; italics in original]

These considerations [considerations that are amazing!—it would be very easy to think they came from a contemporary psychoanalyst, had one not known that they were written around 1930] clarify the phenomena of splitting of the "true self" and its envelopment by the wise and precocious "false self"; all of which is directly linked with Ferenczi's understanding of the reality of trauma. In this way, he bases analysis in the *reality* of the *major points* of impact of a subject's *history*—in his truth.

Indeed, the idea of trauma is the conceptual bridge that links the external reality of the past, the events of life and their consequences on the subject's internal world, to its influence on unconscious expectations, on the crystallization of, and/or increase in, their strength.

It should not be forgotten that the re-evaluation of the concept of trauma, in contemporary psychoanalysis, also marks the re-opening of a very old file—at least as old as psychoanalysis itself. I am referring to *the link between external and internal realities*, between an event and its precipitate, the effect that an event might have on the internal world of the subject. It is a difficult problem, a complex and thorny one that it is often easier to dismiss. But the impact of reality, and especially of events that are justifiably described as "outstanding", upon the internal life and even on the fantasies is undeniable.

For Freud, it is "nonsensical to say that one is practising psychoanalysis if one excludes from examination and consideration precisely these earliest periods" ["*Urzeiten*"] (Freud, 1939a [1937–39], p. 73). In a letter to Ferenczi he writes:

> The outline of your new ideas on the traumatic fragmentation of the psychic life is most stimulating and has something of the general characteristics of "Thalassa". It seems to me, though, that considering the extraordinary synthetic activity of the ego, one cannot really talk about trauma without at the same time dealing with *reactive cicatrization*. The latter engenders what is visible to us—traumas we have to infer. [Fr. to Fer., 16 September 1930]

Inferring them, guessing them—in other words, analysing these sequences of events—becomes a task of paramount importance for the analyst, according to Ferenczi. In a paper that appeared a year after his death, in 1934, in the *Internationale Zeitschrift für Psychoanalyse*, collating five notes written at different times between 1920 and 1932 under the title "Some Thoughts on Trauma", Ferenczi (1934 [296]) reports the dream of a young female patient:

> A young girl (child?) lies at the bottom of a boat white and almost dead. Above her a huge man oppressing her with his face. Behind them in the boat a second man is standing, somebody well known to her, and the girl is ashamed that this man witnesses the event. The boat is surrounded by enormously high, steep mountains, so that nobody can see them from any direction except perhaps from an aeroplane at an enormous distance. [pp. 241–242]

Ferenczi (1934 [296]) comments: "The first part of the dream corresponds to a scene partly well known to us, partly reconstructed from other dream material, in which the patient as a child slides upwards astride the body of her father and with childish curiosity makes all sorts of discovery trips in search of hidden parts of his body, during which both of them enjoy themselves immensely." He continues: "The scene on the deep lake reproduces the sight of the man unable to control himself", associated to "the thought of what people would say if they knew", represented by the second man in the dream. The feeling of being dead and in distress, as expressed in the dream, is completed by the image of "the depth of unconsciousness, which makes the events inaccessible from all directions", at the most perhaps by "an airman flying very very far away, i.e. emotionally uninterested." He remarks further: "Moreover, the mechanism of projection as the result of the narcissistic split is also represented in the displacement of the events from herself onto 'a girl'."

For Ferenczi (1934 [296]), "the therapeutic aim of the dream analysis is the restoration of direct accessibility to the sensory impressions" with the help of a deep trance, which regresses, as it were, "behind the secondary dream, and brings about the re-living of the events of the trauma in the analysis". The quality of the analyst's contact in his situation is of paramount importance to him. It "demands much tact. If the expectations of the patients are not satisfied completely they awaken cross, or explain to us what we ought to have said or done." He adds: "The analyst must swallow a good deal and he must learn to renounce his authority as an omniscient being".

The quotation, in his condensed, telegraphic style, contains some important remarks on the transference authority of the analyst and his true position, and, in relation to the latter, on the importance of recognizing his mistakes. With regard to this, Balint (1934, p. 238) points out that Ferenczi never forgot that psychoanalysis had been invented by a patient, and that the merit of his doctor lay precisely in accepting the patient's guidance and being willing to learn from him a new technique of cure. Nor is it by chance that the word "elasticity" comes from a patient.

We are talking about typical situations where confusion or the (defensive) boredom of the analyst reigns, so that he makes a false move or a slip of the tongue in a session in which the analytic process arouses negative feelings in him. Ferenczi's courage in facing those painful occurrences marked by the failure of libidinal aims and in which destructive forces had gained the upper hand led him to question his own *countertransference*. His *Clinical Diary* (1985 [1932]) is a moving testimony of this.

Once he was able to confront his own trauma, Ferenczi could understand it in others. In order to overcome these traumatic situations of helplessness (Freud, 1950a [1895], p. 318), the mother's *words* are vital. What she says and the understanding of it are a protective function against excitations enabling the subject to work through the trauma. ["Helplessness", according to Stern (1985), is the lack of a "self-regulating other". Trauma is thus provoked not only by an event but also by the communicative atmosphere to which it is linked.] In the analytic situation a similar role is played by the analyst. An example, taken from Ferenczi's *Clinical Diary* (1985 [1932]), seems to illustrate this most aptly:

R.N.'s dream. Former patient Dr. Gx. forces her withered breast into R.N.'s mouth. "It isn't what I need; too big, empty—no milk." The patient feels that this dream fragment is a combination of the unconscious contents of the psyches of the analysand and the analyst. She demands that the analyst should "let himself be submerged", even perhaps fall asleep. The analyst's associations in fact move in the direction of an episode in his infancy ("*száraz dajka* affair", at the age of one year); meanwhile the patient repeats in the dream, scenes of horrifying events at the ages of one and a half, three, five, and eleven and a half, and their interpretation. The analyst is able, for the first time, to link emotions with the above primal event and thus endow that even with the feeling of a real experience. Simultaneously the patient succeeds in gaining insight, far more penetrating than before, into the reality of these events that have been repeated so often on an intellectual level. At her insistence and demand, I help her by asking simple questions that compel her to think. I must address her as if she were a patient in a mental hospital, using her childhood nicknames, and force her to admit to the reality of the facts, in spite of their painful nature. It is as though two halves had combined to form a whole soul. The emotions of the analyst combine with the ideas of the analysand, and the ideas of the analyst (representational images) with the emotions of the analysand; in this way the otherwise lifeless images become events, and the empty emotional tumult acquires an intellectual content. [pp. 13–14]

This suggests a technique not of a river flowing smoothly along but rather one *full of intense, difficult moments linked to traumas* against a general current of the double play of transference and countertransference, a communicative long-term relationship.

Sándor Ferenczi's vision of psychoanalysis is of a place where the demons of trauma can be worked through once they have been re-experienced and re-lived, through the mediation of a *word* that expresses the analyst's *understanding* and acceptance. He *leads* the analysand to *accept himself* in an atmosphere of honesty and genuineness. It is Ferenczi's response to trauma that he himself suffered so intensely.

Attila József, one of his compatriots and a contemporary considered to be the greatest Hungarian poet of the twentieth century, a traumatized figure if ever there was one, has expressed it in a well-known and often-recited poem. This poet was in analysis at the time and describes just such a situation.

> *You have made me the child again*
> *without a trace of thirty years of pain.*
> *I cannot move away,*
> *in all I do, despite myself*
> *it is to you that I am drawn.*
> . . .
> *I slept on the threshold of life*
> *rejected from a mother's arms*
> *hiding within myself, insane.*
> *Above, a vacant heaven,*
> *beneath, insensate stone*
> *Sleep!*
> *It is at your door that I come knocking.*

There are some who weep in silence
but yet are hard like me.
Look: my love for you
is of such strength
that with you now
I also love myself.

Attila József (1936, from the French—Guillevic)

Ferenczi took seriously and pushed to its ultimate conclusion Freud's teaching according to which man is, from birth, endowed with "libido", the aim of which is to create links (Freud, 1923b, p. 45; 1925h, p. 239). Every important failure, brought about by the breakdown of this force, every "catastrophe" (Ferenczi), must have crucial repercussions; this, in a few words, is the meaning of trauma. Is not this the romantic vision of the human being? German romanticism is certainly no stranger to the emphasis Freud and Ferenczi placed on those aspects of man hidden behind a façade of reason. We owe to psychoanalysis the recognition of the central role of affectivity, fantasies, Eros, and Thanatos. The problem of trauma is at the cross-roads of these ideas.

Traumas: are we talking about one, or many? If we take up this question again in the light of what has been said, it would appear useful to recall the *scientific context* in which Ferenczi worked. The author of "Thalassa", fascinated by geology and *palaeontology*—as Freud was by archaeology—could not but be influenced by the theories current at that time, during which the "catastrophistic" theory was predominant, giving place only little by little to the "gradualistic" theory—that is, to an image of evolution in which the determining factor is no longer the single catastrophe but multiple stages. Similarly, in psychoanalytic conceptualization, the model of a vivid traumatic event could prepare the way for the investigation of the complexity of successive events in the history of the subject, for a conception of *multiple traumas* or repetitive traumas. (Masud Khan, 1963, refers to "cumulative" traumas. Freud had, however, already talked about partial traumas ["*Partialtraumen*"] and "part components of a single story of suffering" ["*Stücke einer Leidensgeschichte*"]—Freud, 1895d.)

The professional life of analysts is riddled with such traumas. Ferenczi certainly suffered a trauma in his relationship

with Freud, and that had consequences for the whole psychoanalytic movement. I am thinking of the difficulties experienced
between Freud and Ferenczi, at the end of Ferenczi's life, over
problems of the *technique* and *practice* of psychoanalysis [for a
more detailed discussion, see Haynal, 1987a, especially chap-
ters 1, 2, and 3]. They are not chance occurrences, for they
touch on one of the most delicate problems of psychoanalytic
practice—the repetition of trauma in treatment and its implications for the analyst. But in speaking of trauma with regard to
Ferenczi, it should be clearly stated that in the historical context it was fear for the survival of psychoanalytic doctrine that
overshadowed the freedom of research. The right to disagree
was thrown into question by considerations that, in the end,
cast gloom on the relationship between the pioneers of psychoanalysis. However, there is no doubt at all that the very profound bond between the two men—Freud's nobility of mind
and Ferenczi's attachment to his master—in the end preserved
their relationship. Their solidarity prevailed, and discord did
not degenerate into hostility—which could not be said for other
members of the Committee who found themselves involved in
such tension.

It is important to examine the reasons for this. I return to
the issue in the chapter on the history of the problem.

Working through, mourning, and freeing

It is sometimes forgotten, because of the inevitably morose connotation of the term, that mourning is the prelude to the *freeing*
that results in a new life.

This is a painful process, often accompanied by sadness
and *suffering*. It is also an instrument for mastering the negative aspects of those unremitting changes that the human
being has to undergo throughout his life. As we have seen, the
psychoanalytic process is nourished by forces productive of
hopes and ideas, sometimes disproportionate, that enable the
subject to confront aspects of his personality and relationships
that are problematic for him. In actual fact, it represents a
confrontation with a certain "reality"—this time an internal

one. In the mourning process, the subject rebels against this confrontation and tries to defend himself with the most implausible mechanism: for example, by denying obvious facts such as loss or death (denial) or by trying to push them out of consciousness (repression, splitting). These facts then regain entry into consciousness, presenting themselves as threats. It is in this internal work provoked by the "re-exposing" of painful themes, the main work of the psychoanalytic process, that the perspective of change begins to take shape on the horizon. It involves relinquishing some aspects of the person (his self-representations) or of his relationships, and this is, inevitably, accompanied by anxiety, pain, and sadness. The subject gets into a vicious circle, a kind of a spiral, of wanting to let go while finding it painful. Because of the *suffering* that surfaces again and again, he repeatedly resorts to defence mechanisms, which, in the context of the analytic process, are referred to as *resistances*. The wish for change, however, leads him further afield; the subject is in conflict, torn, rejected, and projected. When he is able to accomplish this process endlessly himself, the analysis will be finished: a good criterion for a possible end to the work with a third person, the analyst.

This *dramatic* process of the *mourning* situation is set in motion by suffering, which demands a withdrawal of cathexis at least in its present form. Lagache said that it consists of killing the dead person [". . . the meaning of the work of mourning is the destruction of the lost love object . . ." (Lagache, 1993)]. It is achieved *under the pressure* of suffering. A forward step is made, followed by a defensive one, in a constant to-and-fro repetition of the spiral, until some arrangement is reached with his "demons" and his "deaths" (Haynal, 1976).

This same process is accompanied by introjections, *replacing* other contents, which until then were present in the personality, in various relational valences (cathexes). From another viewpoint, these are the introjections that constitute memory (Ferenczi, 1909 [67], 1912 [84]). Ferenczi has shown that they are ubiquitous: in everyday life, the "loss" of an experienced moment is introjected in order to retain it: this is the basis of memory. In some moments of *identification*, notably within the psychoanalytic process, this mechanism becomes more apprehensible, in the fraction of an element of mourning and introjection. It is a struggle between affective cathexes forming an intrapsychic conflict: letting go or retaining. The affective price of this relinquishing may well be too high in terms of depression.

We *retain* the present moment in our memory—in Freud's language "*Er-innerung*"—as an expression of the constancy principle, the conservational force of the mind. It is this, therefore, that we relinquish. [The three verbs—"to replace", "to give up" ("to relinquish"), "to retain"—all correspond to the verb "*aufheben*" in German, which has a triple meaning—"to compensate", "to preserve", "to abolish". In writing these paragraphs I used all three meanings successively without being aware of the relationship between them. I am grateful to Ernst Falzeder for having drawn my attention to this interesting fact.] "Retaining" also means acting against the feeling of loss. It is a compensation, sometimes even a triumph, although the loss may be experienced as a release from the painful present (Haynal, 1976, 1987b).

This return to Ferenczi's notion of the ubiquitousness and quasi-continuity of *introjections* (a movement composed of "keeping in the internal world" what has been lost in the external one), the return to the problem of what constitutes memory and the learning of experiences, which is one of the basic characteristics of *every* living organism, is also a meeting point with other contemporary scientific views. "*Er-innerung*", remembering, is a taking-within the self of all that has been experienced, and it is a multiple accumulation. This is what contemporary science teaches, and it responds to Freud's intuition [". . . memory is present not only once but many times, and is composed of all manner of different 'signs'"] (letter to Fliess, 6 December 1896).

A certain mobility in these cognitive elements is ensured through the hope of being able to change some factors in them, a hope reinforced by the analyst's seduction and by expectations projected onto him—a promise of freedom (Pollock, 1977) from earlier cathexes to which the subject had been attached, having procured pleasure from them and because he was used to them, is responsible for the painful element in change, the *mourning* aspect of it. The partly painful "working-through" of the material brings about the resurgence of distressing if not traumatic memories. It is in this working through, during which conflicts re-emerge and traumas are confronted, that the main affective centre of psychoanalysis lies. It is an arduous task but one full of hope and sometimes even amusing.

Etymologically, the word "analysis" (from Greek "*ana*" and "*luo*") refers to a dissolution of rigid structures in the sense of a breaking down into constituent elements. The analysis of fantasies contained in dreams and in the relationship that the analysand forms with his analyst and with the external world is based on the understanding of the constituent elements of these configurations, which relocates them both in their historical and current contexts. The bound affects linked to these configurations partially prevent their being penetrated and understood directly: different defences, such as the externalizing in projection or the dismissing from consciousness through denial and repression, make certain elements disappear, and it is only after a great deal of work *against the resistances* that access can be gained to a more complete version of these configurations. The normal passage of *information* to the inner world of the subject, and between him and his environment, is also hampered by these defence mechanisms, which are triggered off by—and whose *raison d'être* are—anxiety and other negative affects (such as shame, guilt, sadness, depression, and hopelessness).

The powerful *mobilization* of affects that governs psychoanalytic treatment—the cathexis of the analyst, of the setting that he provides (his consulting room, its furniture and lights, his voice and manner, etc.), and of the idea of psychoanalysis that sometimes seems to be most caricatured in the novice analysand—creates *illusions*: the analyst becomes the father, mother, mistress, friend, son, etc., in the various scenarios that the analysand plays and replays with him and which both endeavour to understand through constructions and reconstructions. These illusions will be lost in a process of constant *mourning*—made for the most part unrecognizable by strong defences that it is subjected to (in the analysand, but also in the analyst, in whom traces can be found in the innermost recesses of his theorizing). In its turn this will facilitate the working-through of the controlled loss of illusions: it should be assimilated in part and in part rejected in the working-through in the analytic work. It permeates the whole constitution, and above all the reconstitution and reorganization of the mind in the psychoanalytic treatment. It does so, not only from the viewpoint of representations, but from that of *affect*; under the influence of the "repetition compulsion" (Freud, 1914g, p. 150), the repetition of *representations* and complexities of representations, of scenarios, which we call *fantasies*, also determines a repetition of *affects*, even if part of them is kept at bay, in repression, or split off elsewhere. This "*Erlebnis*" [experience] will be followed by "*Einsicht*" [insight]. In a way, the analyst is a "baby-sitter" who, in the play of profound *affective communication* with the analysand sometimes understands the affects expressed and puts it into words, as a mother figure sitting with the baby would do; at other times he is immersed in the "play", engulfed and sometimes swallowed up in a shared experience from which both of them must subsequently release themselves. And this movement of "release" will in effect be the mourning of illusions and the freeing of the demons behind them, which may sometimes be "death-bringing" because they are born of trauma.

Specific illusions form a very important part of internal object *relationships*. When Freud refers to transference as "our cross" (Freud, 1963a, letter to Pfister, 5 June 1910), it is as a function of the enormous expectations it creates with regard to the analyst, arising from the illusions of activated unconscious fantasies. Obviously, these illusions are only gradually lost in a painful mourning on both sides. For the analysand, it will be the mourning of disproportionate hopes provoking attacks, narcissistic blows, anxieties, and defence mechanisms.

The analytic process starts off *from* these illusions. The analyst helps the analysand to localize the areas of sensitivity, of vulnerability, the affects—especially the negative ones—hidden behind them, and the defences used. *Interpretation* communicates the analyst's empathy ["*Einfühlung*"] and at the same time—hopefully—gives rise to insight.

To my mind, the *discontinuity* of the analysis plays an equally important role (Haynal, 1976), and this goes back to a dimension that has not been sufficiently examined up to now—namely, time. A change of representations means development in time, returning to the idea of process, "an ensemble of phenomena regarded as active and organized in time" (according to "Petit Robert"—Robert, 1988). The word "process" can signify evolution—change in continuity or, conversely, in discontinuity.

The analytic process is, paradoxically, sustained in part by discontinuity, by movements of "digestion" (introjection) represented by the sessions, episodes, experiences, and understanding gained in the psychoanalytic encounter. Discontinuity is the creation of dreams after the day's interruption; it is the internalization of what has happened in the outside world. It is thus the creation of memory: the images we retain of identifications, whether or not they seem important; and it is the progressive loss of illusions during analysis, their working-through, accommodating to our demons and the adjustment of the images to each other. In other words, it is the whole of internal life and the whole *process* of psychoanalysis. What Pontalis (1978) describes as "progressive absence" results in an end to the analytic relationship. Furthermore, these discontinuities belong to every modern theory of nature up to R. Thom's (1974) theory of catastrophies.

The first time I presented my ideas on this link with mourning (Haynal, 1976), there were some reservations that they represented a pessimistic view of psychoanalysis, even an attempt to swallow it up in a moment of mourning, thus making it a sad affair. I am inclined to see this as an affective resistance to recognizing that *losing* is a basic fact of human life, of successive generations, of mobility, renewal, and the source of creativity (Haynal, 1987b). Destruction and the recreation that follows —i.e. "reparation"—return to inevitable losses as sources of creative-reparative forces, a process I described in a group of orphans with a very high degree of creativity that at times was almost extraordinary (Haynal, 1978).

So let us remember that psychoanalytic communication is not only the communication of representations, of a fantasy, a text; this *communication through affect*, unconscious and non-verbal, or "from one unconscious to the other" ["It is a very remarkable thing that the Ucs of one human being can react upon that of another, without passing through the Cs . . . descriptively speaking the fact is incontestable" (Freud, 1915e, p. 194)] takes place through minimal signs linked to the tone of voice, the handshake, a whole range of bodily expressions of affect. From the start, psychoanalysts such as Ferenczi and Reich, the last leader of the technique seminar at the Vienna Psychoanalytic Institute between 1924 and 1930, attributed an enormous importance to it, though recently it has been chased from the field of interest in favour of a purely verbal "textual" model. And yet, in psychoanalytic technique, understanding is achieved not only through insight [*"Einsicht"*], but also through experience [*"Erlebnis"*] (Ferenczi, 1924 [264], pp. 7–8); affect, therefore has a part to play as well as the text. The "play of language" involves transmission and sensitivity to the subtleties conveyed.

The affects in question cover the whole spectrum, from joy, surprise, and interest to the negative affects, with which psychoanalysis is more particularly concerned. Although the role of the latter is incontestable in pathology and a source of suffering, which is what brings subjects into analysis, nevertheless the discovery of its importance was not without difficulties in the course of history. The central role of anxiety overshadowed the depressive affect (a factor about which Abraham was already complaining in 1911); other negative affects, such as scorn, hatred, hatred of the self, received sustained attention only in post-Freudian developments, notable by Melanie Klein and W. Bion. The importance of the renewal of our theory of affect is dealt with in chapter four, which is devoted to that particular problem.

. . . THE *WIDE-ANGLE* VIEW

CHAPTER TWO

Meaning

> . . . the real understanding of an idea or a theory
> involves the subject's re-invention of it. . . .
>
> Piaget, 1973, translated from the French

One of the most important concepts in the psychoanalytic encounter is *meaning*, man being a creature who is constantly creating meaning. "Meaning" is different from "knowing" and different again from "science". Defined in the *Concise Oxford Dictionary* as "1. what is meant by a word, idea, action etc. 2. significance. 3. importance", it is, in the final analysis, a *semantic field* (an ensemble of representations) created by vocalization, gesture, posture, etc. and the perspective in which our internal and external perceptions are *interpreted*. The meaning *attributed* to the perception of objects, persons, events, scenarios can engender conflict and suffering. Epictetus already declared in the *Enchiridion* (p. 79) that men are troubled not by things but by the opinions they have of them, and Shakespeare says in "Hamlet":

> . . . there is nothing either good or bad but thinking makes
> it so. [Act II]

So there is a wish to revise these perspectives and to *change their meaning*, and thus change the representations—or some of them—linked to the events. This can happen in the setting of a privileged encounter like psychoanalysis.

Another concept essential to our investigation is that of *communication*. In a situation like psychoanalysis, characterized by the absence of a fixed and obvious "purposive idea" [Freud's concept of *"Zielvorstellung"* (1990a, p. 530)], communication enables the other to understand. The person *communicates*, and what interest us in the communication are (1) his emotions, (2) the ideational—"cognitive"—constellations that they mobilize, and (3) the habitual modes of mastering them, of "defending" himself. There is no need, therefore, to point out the importance of communication, both verbal and non-verbal, except to say that, in the communication of emotions particularly, the non-verbal part, which is not immediately conscious, is of major interest. A *wide range* of various positive and negative affective states are put into play in communication, going beyond one single anxiety or anxiety coupled with depressive affect (even though there is no question of the importance of these two frequently studied emotions—a theoretical importance, especially with regard to the mobilizing of defence mechanisms).

Among the cognitive constellations, some of which are unconscious, there appear "Gestalts" that we call *"fantasies"* which often translate instinctual needs, *wishes*—especially sexual ones in the broad sense of the term as Freud defined it (Freud, 1900a, pp. 565, 598). The analyst's function will be not only in reading the emotions, fantasies, and ways of mastering the situation as characterized by their emergence, but also in *communicating* what he has read. And let us not forget that communication is *action*. It is more than a re-description; it contains, perforce, assertive elements (affirmation of the way the analyst has understood the communication), and directive elements (the implicit discouragement to act out: cancelling a session, for instance, ill-treating a child, etc.). There are also elements of engagement (evidence, for example, that the analyst is continuing to listen attentively), elements of expression (the analyst's state of mind is manifest in the sound of his voice and other non-verbal signs), and, finally, elements of declaration ("we must stop now"). These are categories suggested by Searle (1969). The alleged neutrality, or, in Freud's terminology, "indifference", implies an emotional distance—a distance taken by the analyst for reflecting on his own emotions—that is, his internal work. But this mastering of his emotional reactions does not imply that he can *avoid experiencing and participating in the communication* according to the rules of all communication. Incidentally, Freud never used the term "neutrality" in this context in any of his published work. He only used it in another sense and in a private context in a letter to Ferenczi (26 February 1916): ". . . All your communications have been of great interest to me. But you will allow me to maintain a benevolent neutrality, partly because I am too lazy and partly because I would not want to upset your efforts" [". . . *Alle Ihre Mitteilungen haben mich ungemein interessiert. Sie erlauben mir aber eine wohlwollende Neutralität zu beobachten, da ich einerseits zu faul bin anderseits Ihre Anläufe nicht stören möchte"*]. He knew the term but never used it in a technical sense. Freud's remark (1915a, p. 164) about the term *"Indifferenz"*, which Strachey translated as "neutrality", poses a problem, since as a matter of fact Freud was talking about indifference and not about neutrality. Strachey corrected Freud, but Freud was not corrected in this particular case, although he had often added remarks elsewhere in his texts to point out what in the meantime seemed to him outdated (for example: "I am aware that no analyst can read this case history today without a smile of pity . . ." etc.—1895d, p. 105, additional note 1924).

Furthermore, in *reconstructing* what he has heard and experienced, he will inevitably interpret (in the original sense of the word and in this case not in the technical sense of the word). He *understands* what he has heard in the perspective of what he knows about the individual's history and functioning (the dichronic and synchronic aspects): his new understanding and the communication of it may offer to the analysand perspectives that had not occurred to him previously, thereby making "conscious" what, until then, had not been available.

The psychoanalyst knows—because he *remembers*—a more *global context* than the analysand keeps in mind at any one moment, so making available to the subject a "narrative", im-

ages in which those elements are integrated. In his remembering, the analyst does not have the same personal reasons as
does the analysand to push them away (by repression or splitting, for example).

The concepts of "scenario" (Schank, 1972; Schank &
Abelson, 1977), "setting" (Minsky, 1975), and other similar notions help us to pursue an understanding of the function of
interpretation. For instance, in the analyst who is listening to
the account of a 39-year-old woman, it gradually becomes apparent that he has already heard a similar constellation in her
life—for instance from her childhood. For Juliette finding herself in a situation with two women, her mother and her nurse,
is not new. The one is respected and distant; the other offers
the sensual and a feeling of security. In talking about the
present, she recreates this situation with its same characteristics—without realizing it. The analyst knows the scenario. . . .
His "cultural advantage", defined by his position in the external
world, enables him to group the social conventions more per-
spicaciously than would someone not outside, who, through
his "place" in the social field, is less inclined to wonder about
them. [They do not give to the observer the security that they
would to people who have not questioned the value of the idea
concerned, and the observer is therefore less identified with
them.] His personal analysis, too, represents an "advantage" in
the experience of challenging conventions and the beneficial
effects resulting from them. It enables him, through "counteridentification", to further a similar challenge in the other.

It is clear that in order to analyse the contents of the analysand, the psychoanalyst has to be able to *communicate* them to
him. The problems of the psychoanalytic communication and
questions related to its possibilities lie at the *centre* of every
discussion on analytic practice.

CHAPTER THREE

Communication

I f thinking is verifiable only on condition that it is immersed in empirical practice, psychoanalytic thinking can only be based on the practice of analysis. At the same time it may include aspects of communication in the psychoanalytic situation and clarify some of the structures of all interpersonal communication: the conceptual grasp of this reality from Freud to Bion is ample demonstration. If *reason* is only a working tool, if its *fallibility* is acknowledged as Freud did many a time when he talked about a provisional (Freud, 1900a, p. 281; 1905d, p. 242; 1912–13, p. 97; 1940a [1938], p. 159) theoretical framework (Freud 1900a, pp. 536, 568, 598, 610), truth can no longer, today, be considered immutable. It is, rather, an ensemble of verbal and non-verbal games that gives us a *hold* over the world—no mean feat. In that sense, straddling modernity and post-modernity, Freud opened the way to post-modernity— modern, because despite his uncertain epistemological positions and his referral to positivist models, he did work his way towards the construction of instruments likely to extend the influence of reason in the to-and-fro between theory and practice. In doing so, he went beyond contemporary philosophers while taking up their requirements.

If pragmatism is none other than the study of relations between speech acts and the situations in which they are produced, psychoanalysis has created an exemplary framework for the exploration of it. Psychoanalytic theory provides a perspective for the scientific study of the phenomena of communication—social communication—and the functioning of reason (the ego).

According to Habermas (1985), the meaning and aim of the "discourse"—that is, of the dialogue, of argumentations—is the search for a consensus, a negotiation between *equilibrium* and *innovation*. These elements must, of course, be present in every psychoanalysis, especially if it is seen from the viewpoint of *emancipating* the individual. Focusing on the subject in a communication process, which is characteristic of psychoanalysis, comes close to post-modernist ideals. Habermas (1985), leaning on other authors, defines modernity as "a *contact* which has become *reflected*, having lost its naturalness through the universalization of norms of behaviour and modes of socialization . . . which force individuation. . . ." In this context and the context of *post-modernity*, psychoanalysis may be the object of an apragmatic reflection. If it is acknowledged that subjectivity determines its fulfilment from structures that are characteristic of it and that in psychoanalysis it is the mainspring of a process leading to self-fulfilment, on account of the analysis that it makes of communication processes, it becomes an instrument of *reconciliation* between the subject and himself.

Study of the communicational dimension in psychoanalysis facilitates the move towards a theory of rationality, a theory of links, and a theory of modernity in which those aspects offered by Freudian traditions are assimilated and become an integral part of man's reflection on himself.

An intersubjective approach is, of necessity, one capable of restoring the complete dimension of verbal and non-verbal language in an observable situation in which questioning is possible. But this intersubjectivity may itself be questioned and, on a "meta" level, understood like points of condensation, as in Freudian theory. It also implies renewing that theory in the light of today's questioning, which is easier to accomplish

against a background of understanding of the interwoven relationships between its original form and the culture of the day.
That is why it is important to sketch a history of psychoanalytic
ideas as they interact with their surrounding culture. I focus on
this in part two of the book.

Psychoanalysis is a *particular kind of communicative practice* centred on the exploration of *one* of the protagonists in the
encounter and of the other in so far as his activity demands it.
(The necessity for this, incidentally is subject to controversy:
see Haynal, 1987a). My thinking is focused upon the communicative practice as it appears in the analytic situation, including
problems of the relationship of verbal to non-verbal, the function of the word in individual homeostasis, and thereby the
expression and mastery of emotions in that situation. Problems
in the interaction both of communication and body functions in
these processes may be touched on from the point of view of
interaction through voice and gesture. The analysis of posture
should not be neglected either, despite the fact that as a result
of historical development, probably linked to difficulties with
regard to Wilhelm Reich's way of thinking, the literature offers
nothing original beyond the widely known clinical experience
shared by practitioners in the field.

For the analysand to be able to *communicate* what he contains, the communication has to be *received* through some process of *identification*, in the broad sense of the term, not only
with what the analyst (consciously) understands, but also with
what he feels only "vaguely" (unconsciously or preconsciously),
in his *countertransference*. In this way he soaks up the narrative, the slips of the tongue, parapraxes, dreams, the sound of
the voice and its tone, behaviour—in short, the whole semiology
of communication.

"*Neutrality*" is an asymptotic position: the psychoanalyst
aims at *recapturing* the internal attitude that Freud referred to
in the expression "*Niederhaltung der Gegenübertragung*" ["the
taming of the countertransference"—in German the literal
translation would be "the holding-down" of the countertransference (Freud, 1915a, p. 165)]. It implies that he will not react
in accordance with his emotions and desires but will try to
understand what is going on within himself, thus tending to
take up again and again his position as analyst. The clearest

expression of this opinion is contained in Binswanger's memoirs: "Each time, therefore, one must recognize and then go beyond one's countertransference; only then can there be freedom within oneself" (Binswanger, 1956, p. 65). The other expression that Freud liked is "*Überwindung*" [overcoming—it also means "willpower"): "to overcome the countertransference" (Freud, 1961a, letter to Jung, 2 February 1910), "complete surpassing" (Nunberg & Federn, 1967, p. 437). And in another letter to Jung (7 June 1909) he writes: "One needs the strength to master it." [". . . *nötige harte Haut, man wird der Gegenübertragung Herr*"—"with a thick-enough skin one can become master of the countertransference"].

There is some exciting research in this area, particularly Freud's and Ferenczi's on the phenomena of telepathy [*"Gedankenübertragung"*: thought *transference*]. The overlapping between them is such that their reports of communication through processes already set in motion are continued even at a distance, with similar thoughts or "intuitions" punctuating the reports.

It is worth noting that Freud's notion of the unconscious (like C. G. Jung's) was inspired by German romanticism. His research therefore aimed at finding some substance, some essence, something apprehensible hidden in the darker reaches, like the drives or other mythical beings. Freud's epistemology, linked as it was to his scientific training, enabled him to avoid the vitalist trap by "dialectizing" the problem. With the help of "border-line concepts" like drive, for example, he avoided any kind of idealism or mysticism that would have led to the impasse that was Jung's destiny.

Contemporary views endeavour to understand *how* interactions between the subject and his environment and defensive movements (such as "keeping something out of consciousness") "*create*" elements of the "unconscious". Understanding this process by means of contemporary scientific concepts such as communication is not an easy task. The fear, indeed the accusation, of an unacceptable *revisionism*, is not, to my mind, justified. Bringing the understanding of a particular area up to date in the light of other sciences is a useful scientific procedure in which interdisciplinarity and cross-fertilization contribute to increase intelligibility. In this way, the unknown becomes part of a schema of contemporary knowledge, without in any way abandoning what is *specific* in a given field.

Pondering about the start of communication addressed to the analysand, Freud writes: "Not until an effective transference has been established . . . a proper *rapport* with him" (Freud, 1913c, p. 139). (The word "rapport" is used in the 1913 text, thus returning to the terminology of hypnosis. The idea of discontinuity and ruptures—epistemological and others—demands a revision of the most serious kind!) It means that such communication begins when there exists an *affective mobilizing relationship*. Clearly, this relationship outside communication, which is at the level of insight—that is, of understanding ("cognitive" mechanism)—is also a profound affective communication.

Subsequently, in his article on "Psychoanalysis" written for an encyclopaedia, Freud qualifies his new technique of "the art of interpretation" ["*Deutungskunst*"]. He emphasizes that "it altered the picture of the treatment so greatly, brought the physician into such a new relation to the patient and produced so many surprising results . . .", and that the analyst should "surrender his own unconscious mental activity . . ." so as to be able to "catch the drift of the patient's unconscious with his own unconscious" (Freud, 1923a [1922], pp. 238–239).

Freud has produced a superb description of a new form of professional communication based on a deep level of sensitivity, the psychoanalyst being almost impregnated with the impressions he receives without any strain likely to upset the process.

Viewed in this light, psychoanalysis is undoubtedly not merely an exchange of *representations*—as it almost is with Jacques Lacan, where the analysand pours out a text and the analyst gives his associations, his understanding, and his interpretation of that "text". Rather, it is a communication through affect, in which notions such as affect, drive, object relationship, transference/countertransference facilitate efforts at grasping what goes on beyond the exchange of representations. Because the idea of fantasy covers a particular, and a particularly *pregnant*, constellation of *representations*, and because interpretation reflects an *understanding* of those representations (including interpretations that the subject has himself already made for himself) in the representation–affect diad, the part of *affect* must maintain its fundamental primacy. It is one of the stakes in the historic discussion between Freud and Ferenczi: this importance of experience ["*Erlebnis*"] over and above that of insight ["*Einsicht*"] and understanding, and the power of experience to lead to a better insight.

In traditional terms, it is an *object relationship*, which permits an understanding of psychoanalysis as an emotional *experience* between two people: the analyst and the analysand. As is known, Freud emphasized that the emotions that are mobilized in the analysis are *true* emotions: ". . . can we truly say that the state of being in love which becomes manifest in analytic treatment is not a real love?" ["eine *wahre* Liebe"] (Freud, 1915a, p. 168).

This leads to a conception of psychoanalysis in which the analyst *identifies* and brings to the fore those elements from which the unconscious is built—in other words, the themes of the individual's life. As a result, the analysis does not "liquidate" the problems but identifies them and produces a theme for a possible subsequent *working-through* during the course of a life in which the power of repetition will perforce play a role—hopefully a less pathological role, creating less suffering than before, through sublimation, reaction formations, and creativity.

Such a conception puts a firm emphasis on the *emotional* experience of the analysand and on the *repetition* of his experiences ["*Erlebnisse*"], including that of *trauma*, as I have already emphasized. What is specific to psychoanalysis is the *analyst's position*: he is present "*in the name* of" the unconscious, or—to quote Bion—with his "faith" in the unconscious. He perceives and hears what he hears, what he feels, from the *standpoint* of an unconscious: that is the *difference* between analytic listening and other kinds of listening.

Over and above the fact that interpretation is likely to help the analysand, it also serves as a means of *freeing* the analyst. This freeing process has a general importance in psychoanalysis. By its very framework, analysis *constructs illusions* (for example, the analyst is the father or mother, he is fierce or in love, he is especially gentle or favours the patient . . .). These illusions—the transference, the *chimera* of the past—are necessary, but it is the *release* from them through interpretation, the work of *mourning*, the loss of those illusions, that is the *essence* of psychoanalytic "work" (the psychoanalytic *process*) (cf. Haynal, 1987b).

Another question arises: what happens *afterwards* to each of the protagonists? Having mobilized his feelings, the analyst will also be attached to the analysand. And what happens to the *analysand* in relation to those same forces? It seems simpler because it can be acknowledged that the transference will diminish little by little through understanding: through interpretation and "*Einsicht*". But does it really *disappear* altogether?

In this process, an object from the *past* is projected onto the present object. It is not so easy to acknowledge this aspect of the past without its possibly being felt as something "thrown up" from a painful past that it would be preferable to forget. Every perception, however, contains some characteristics of past objects. The object is therefore perceived in a more or less distorted fashion, depending on individual sensitivity. In other words, the degree of mobilization—assuredly unconscious—of the past object will colour certain aspects of the present object. All of this results in a grasp of the *eternal* character of transference: the mother or other significant objects from the past will always be present in the beloved one, even if psychoanalysis can in some way "purify" the perceptions, distinguish them, permit nuances. In the final resort, emotionally every so-called objective perception is in fact an illusion: two people listening to Mozart will not hear the same music.

No matter what the object perception, some part of it is always a projection, and in analysis this applies *a fortiori* to the analyst. He feels the role that he is expected to play (see Sandler, 1976) through the analysand's "illocutory discourse"—in other words, through the importance of the analysand's projection and projective identification on the one hand, and through his own projective counteridentification on the other.

But is this all that occurs emotionally between analyst and analysand? Balint's idea of a "new beginning" goes back to another perspective. The totality of the situation is not determined by a pure repetition, but after a good working-through new contents, new elements will come into play, a new combination: a creative integration is born.

Ferenczi put the problem of "*tact*" at the centre of technical considerations. "Tact" refers to a certain way of "touching". Empathy enables the situation to be understood through a trial identification. Tact describes *how* "to touch" the Other, bearing in mind what it means in the perspective of the past: a maternal or a paternal touch, linked to this or that situation, etc., determines the way (the "how") of giving this information.

Since in our understanding of the other we use our own instruments of internal perception and sensitivity, it is important that we maintain contact with those perceptions and feelings. Expressions such as "tact" and "*Einfühlung*" (sometimes translated as "empathy", sometimes as "affective harmony") are terms close to experience. They provide us with direct bridges to what we experience personally—a seduction, for instance. If every analysis begins with a seduction, with the mobilizing of the other to create a link, the term "therapeutic alliance" is a highly abstract and value-laden idea for describing such a move.

In the present discussions about the Strachey translations and various tendencies aimed at giving psychoanalysis scientific respectability it is important to remember the tradition initiated by Sigmund Freud, who designated facts and expressions in a form *close to experience*. The same path is being taken by recent critics of ego-psychology, which is turning its attention to the results of sublimations in the ego, whereas Freud was interested in the experiential and instinctual roots. As we have seen, Freud talked about indifference. Experience enables us to question what *degree* of *indifference* (and not of neutrality in the post-Freudian sense, as if this could be a state acquired once and for all regardless of its affective implication), of disinvolvement, is *necessary* to our work, as Freud did (letter to Ferenczi, 13 December 1931), and what degree of it is *inadmissible*. The best-intentioned interventions will, if they express the analyst's wishes and not the analysand's, be fruitless or sometimes even have an opposite effect to the desired one. (I am thinking of a patient who locked up her child during an analytic hour, and another who brought her baby to the session under different pretexts. The two situations were only resolved *after* interpretations and *working-through*.) In talking about psychoanalysis it is important to speak the language of *experience*, of communication, and its most important vehicle (by which it is conveyed): *the affects*.

"... hmm ... hmm ..."

Ferenczi, excellent clinician that he was, wrote in his *Clinical Diary* (1985 [1932]):

> Subtle, barely discernible differences in the handshake, the absence of colour or interest in the voice, the quality of our alertness or inertia in following and responding to what the patient brings up: all these and a hundred other signs allow the patient to guess a great deal about our mood and our feelings. Some maintain with great certainty that they can also perceive our thoughts and feelings quite independently of any outward sign, and even at a distance. [pp. 35–36]

Through his "... hmm ... hmm ..." the psychoanalyst expresses his presence, his state of awareness, a certain interest; through the tone of his voice, its rise and fall, he can also communicate: "That's interesting." ... "I am puzzled by that." ... "Do go on." ... "Is that so? Really? Is that what you think?" There may be good reasons for not articulating these contents, but only suggesting their outline. Indeed, analysts have a long and refined tradition of expressing all manner of subtleties by the sound of their voice and its intonation.

Non-verbal communication is expressed through the dampness of the hand, the pallor of the face, scratching, the position on the couch, smell, and many other signs. It is impossible not to take as communication the bad odour of a female patient who, by neglecting her own body so badly as to arouse my disgust, wants me to feel her own disgust with the world and herself. It is only by talking about it that one can touch this object of communication, which, indeed, is disturbing, so deeply unconscious is it and therefore meaningful.

> One has a vision of the successful end of an analysis, which would be quite similar to the parting of two happy companions who, after years of hard work together have become

friends, but who must realize without any tragic scenes, that life does not consist solely of school friendships, and that each must go on developing according to his own plans for the future. This is how the happy outcome of the parent–child relationship might be imagined. [Ferenczi, 1985 [1932], p. 37]

The analysand's verbal and contextual communication is brought to light through the analyst's *reflecting* whatever he has understood of that communication (and, as I have shown, *communicating*, despite his professional reserve, some of his opinions and feelings on the subject). What is special about the analytic situation is that attention is centred on the analysand's communication: not only on the lexical and semiotic aspects, but on all dimensions that can be grasped by both protagonists of this process. Recognition of aspects that were unconscious up to then *modifies* the images of self and other, just as typical interactions between those two people do—the scripts, the "agenda" called fantasies, which seem to be the most important factors in change.

Studying the semiology of psychoanalytic communication opens the way to communication in human life—in the child, the adult, between lovers, across generations, in society and in private, in daily life and in passions, indeed in madness. Surely this is an area that up to now has made insufficient use of psychoanalytic experience? One of the most important aspects is that of affects, which is the subject of chapter four.

CHAPTER FOUR

Affect

> . . . although we are also ignorant of what an affect is.
>
> Freud, 1926d, p. 132

One of the inadequacies of the psychoanalytic theory of affectivity resides in the fact that, perhaps for good reasons connected with clinical experience, it has been centred, from its inception, on *anxiety* (Freud, 1950a [1887–1902], letter to Fliess, 14 November 1897), while other colours on the affective palette have long remained in the dark. Freud later turned his attention to the *depressive* affect (Freud, 1915b, 1916a, 1917e [1915]) but, while recognizing the importance of affectivity in general, he undoubtedly remained *dissatisfied* with the state of development and knowledge about this subject (e.g. Freud, 1926d [1925], p. 132; 1933a, p. 94, etc.; for the historical development of this see Haynal, 1987a). Apart from the affects mentioned and the erotico-sexual sentiments observed in "hysterics" in Freud's early descriptions, there remains a large number of emotions that have not been mentioned or, if they have—Socarides (1977) names boredom, enthusiasm,

59

confidence, sarcasm, nostalgia, and horror, among others—
have not entered the *conceptual* framework of psychoanalysis.
Because of this they have tended not to engage our attention
(e.g. Bion on arrogance, 1957; Haynal on boredom, 1976). A
wide area of positive affects, therefore, has been insufficiently
appreciated in the psychoanalytic literature. In life, as in psychoanalysis, they obviously play a role that it may be important
to recognize in the complex interplay of *emotional exchanges*.

Anxiety certainly has a central place, even if, in the process
of release from the defence mechanisms, unpleasure is added,
and consequently affects with a negative colouring (e.g. the
threat of a great sadness, disgust).

A number of scientific non-psychoanalytic propositions in
the *general* literature take as their point of departure Tomkins'
work (1984), which lists anxiety, distress, disgust, scorn,
anger, joy, interest (and surprise), shame, and guilt. Although
not exhaustive, this list contains affects, independent of cultural variation, that we have in common with the primates—
thus constituting our phylogenetic heritage (Ekman, 1973).

This brings us to the question of the most basic (or "primary") affects. It can be said that the closest in the phylogenetic programme fundamental to man belong to the first
category: attachment and its derivatives such as love, or everything that serves to protect the self, or again, those that signal
danger. These are the most important elements in a series in
which finally one comes upon the Freudian Eros and Thanatos
and the depressive affect. They all function according to the
pleasure and unpleasure principles—that is, the *hedonistic*
principle, a fundamental quality of affects. In remaining close
to his affects, man thereby remains close to his pleasure (hedonistic pleasure) and avoids unpleasure.

If the existence of primary affects is acknowledged, "derived" affects, the secondary ones, would be determined by
semantic nuances of circumstances and of intensity, such as
the pain–anxiety–fear–terror series.

In academic research on affectivity, the predominant model
has, until recently, been the one of the *arousal* function of
circumstances: thus, for joy, pleasant circumstances, and for
rage, annoying circumstances (Schachter & Singer, 1962). This
hypothesis has, however, yielded little by little to a theory im-

plying a specificity of arousal, i.e. broad categories of emotions where rage, for example, would have a different specific arousal from joy. Within these broad divisions a cognitive complexity is produced as a result of a *semantic differentiation*. Thus, one speaks of anxiety when the source of danger is not clearly identified, of fright when the stimulation is sudden and considerable, of fear when there is a "real" danger, and so on. These various states are probably endowed with a biological substructure that is fundamentally different from other dimensions—joy or happiness, for instance.

An affect is therefore, at one and the same time, a *physiological* modification, a visible expression (*a communication*), and a *subjective* experience. This expression is preceded by an evaluation of the situation—sometimes this is rapid and unconscious, grasping corresponding configurations from the viewpoint of danger to the organism or the individual or, conversely, from the viewpoint of advantages (at the level of elementary needs and wishes). The emotions present themselves, in their turn, as a task: they must be discharged or mastered through different defence mechanisms.

Recent research, particularly Paul Ekman's, has shown clearly that the expression of affects by *gesticulation* is not linked specifically to a culture. In cultures as different as Guinea, Sumatra, Java, and Japan, and in our own civilization, people empathize with photos of angry or disgusted individuals, psychic contents being linked to anger or disgust and not, for example, to joy. The vocabulary (verbal expression) translating the degree of *awareness* of affect and *coping* mechanisms, of defence mechanisms in general, in the psychoanalytic sense of the term, is different in different circumstances, just as are the rules of their *manifestations*. (In some cultures, for example, the suffering connected with grief is expressed to a greater or lesser degree; others are more inhibited, the rules for showing affects being more restrictive.)

Curiously, psychoanalysis itself has never offered a coherent and *global theory* of affectivity (despite a few important attempts: Rapaport, 1953, or Green, 1973). It has never expressed an opinion on the circumstances in which affectivity is set in motion or on the non-verbal and cognitive–verbal processes that are involved. In fact, it does not present any complete theory of affective regulation, despite the grasp of certain mechanisms such as repression, splitting, and other defences that serve to regulate information linked to affectivity and are therefore directly or indirectly regulatory mechanisms of affectivity. Is this because, through its very methodology, psychoanalysis is capable of examining only *partial* aspects of this problem, others being extraneous to the psychoanalytic situation? This is undoubtedly the reason why John Bowlby (1969, 1973, 1980) and Mary Ainsworth and colleagues (1978) were obliged to gather their data outside the analytic situation, realizing that a complete theory of affectivity cannot be a psychoanalytic theory of affectivity, although psychoanalysis has important contributions to make to this subject, notably on the regulation of affects, the conditions that give rise to them, and their communication. Besides, theories derived from experimental facts can be complementary to psychoanalysis, more or less close and quite compatible with it, but they cannot replace psychoanalytic reflection.

In the definition of psychoanalytic theory, the conscious is linked to the verbal, to language ". . . the conscious representation comprises the representation of the thing plus the representation of the word belonging to it, while the unconscious representation is the representation of the thing alone" (Freud, 1915e, p. 201). Unconscious communication is, by definition, *non-verbal.* In the *affective* communication underlying non-verbal communication, the aim is to create links, to regulate closeness or distance, to arouse in the Other feelings that correspond to the needs and wishes of the individual at a given moment. If, in 1926, Freud could say that we do not know what an affect is (1926d [1925], p. 132), he nevertheless located the affects in relation to sexuality, adding that they are in the service of the great destinies of the drives: to create links in the first place through Eros and to create distance, removal, extinction through Thanatos. When Aristotle says that "man, by his nature, is a social animal" (*Politika, I*, 2), he is expressing the same sentiment. Creating links and the movements that introduce them (seduction) also seems to be, for psychoanalysis, one of the most *fundamental* characteristics of man. What Freud expressed in mythological language, modern science discovers in a detailed research of correlations of data and their interpretation (for example, in terms of closeness and distance). Freud's intuition that the affects serve the great aims of the species and of the individual—programmed biologically because he brought in the "drives"—seems to me to remain an essential fact. At least such a view can take account of phenomena we observe in our encounters. Freud's choosing to depend on a "borderline" concept like drive [". . . we cannot help regarding the term 'instinct' as a concept on the frontier between the spheres of psychology and biology" (Freud, 1913j, p. 182); "Instinct appears to us a concept on the frontier between the mental and the somatic . . ." (Freud, 1915c, pp. 121–122)], in fact, takes man's intentionality (this is once again at the centre of discussions in models of artificial intelligence—or, as is sometimes the case, is omitted, at great cost, from them) and re-links this to the fundamental biological processes at the root of human existence. If in psychoanalysis the patient's past is the reference point for the interpretation of his discourse, it is a past composed of the totality of his experiences, including his instinctual "destiny". To forget this would mean no longer practising psychoanalysis in the Freudian tradition—in other words, it would, to my mind, mean *losing* one dimension of the fundamental perspective.

Affect is the result of a modification in a physiological state (involving regulations in the autonomous nervous system and humoral regulations: hormones, neurotransmitters, etc.—Pribram, 1984, and Whybrow, 1984), and, in correlation with this, of a change of attitudes in communication (gesticulations, voice, gesture), of modifications in behaviour or preparatory to such modifications (muscular tension, agitation, etc.), that the subject perceives. These modifications are preceded and followed by a cognitive change triggered by conscious and unconscious processes. Gesticulatory and vocal expressions have been the subject of recent studies (Ekmann, 1980; Goldbeck, Tolkmitt, & Scherer, 1988; Scherer, 1982). There is absolutely no doubt that the affective system, which is manifested in the first instance quite clearly by facial expression, is a phylogenetic heritage (Plutchik, 1980). The preliminary phylogenetic conditions seem to be a certain height and body weight, the development of an upright walk, mainly diurnal habits, and a social context in which such a sign system is an advantage.

Certain modifications to the autonomic nervous system are linked specifically to certain affects and are transcultural and universal—for example, the acceleration of cardiac activity in states of fear and anger and the diminution of it in states of aversion; skin temperature, measured on the fingers, increases in anger and decreases in fear. These data, highlighted by systematic experimentation (Ekman and his school), draw the analyst's attention to a whole range of affective states and to the fact that those states are linked to specific bodily modifications. In other words, in dealing with affects one is in a limitrophic zone between body and mind (which paves the way to a psychoanalytic understanding of psychosomatic phenomena).

Whatever the reality of the *drive* concept as it figures in the psychoanalytic tradition, the very problematic nature of it cannot simply be put aside. In other words, the problems cannot be resolved by dismissing the concept, as some suggest nowadays. First of all, the work of the drives is distinguishable on the clinical level. For example, in periods of sexual abstinence, sexual fantasies are more active than at times when the individual (or the analysand) is sexually satisfied. Analogies with hunger phenomena have been remarked on by Freud and followed up in their action in dreams (see his daughter Anna's dream about strawberries: Freud, 1900a, p. 130; 1901a, p. 644).

It is quite clear that Freud intended the link with the drives to ensure a firm biological footing for his theory in the Darwinian tradition to which he belonged. At a time when studies of animal behaviour through ethology have provided quantities of data for consideration with a view towards a better understanding of man (e.g. Goodall, 1986), this problem does not seem to me to be outmoded.

The drives—let us use this concept for now—are regulatory systems of pathological processes, some of which depend on external objects (the nutritional system or sexuality). As a result of this *contingency*, this dependency on the external object, such acts may become conscious (at the point when the object is lacking): this is the underlying meaning of the birth of awareness through the lack of the object (Freud, 1900a, p. 566). The processes of quenching thirst, satisfying hunger, storing heat and light, absorbing oxygen, and all aspects of eliminating are instinctual processes that can become *conscious*. In the course of a child's development, during the processes of learning, they are modified by reactivations of needs and desires, by information coming from without, by expectations, and by the complex interplay of all these representations. The wish for warmth and the presence of another, along with the functions involved in feeding and evacuation, will be combined in *a system of desire* characteristic of *man*. It is a system that has its point of departure in the most primitive needs as conceptualized according to classical psychoanalytic theory in the area of "orality" (Freud, 1905d) and according to the more recent theories in the area of "attachment" (Bowlby, 1969, 1973, 1980; Lebovici, 1983; Stern, 1985). The representation of the "presence" of the Other, therefore, is confused with the sense of well-being linked to feeding, warmth, the sensation of cleanliness (and, on a more general level, with self-perceptive feelings about the skin), and it is activated by part systems of sexuality.

To eliminate the theory and do without the richness of observations connected to the instinctual system because of their complexity would amount to intellectual self-mutilation. On the other hand, to reconsider the theory of affectivity in the light of a drive theory, taking into account contemporary biological and physiological knowledge, seems to me to be a necessity that should not be bypassed. The drives are one of the *mobilizing* systems of the affects. The encounters—the relational aspects, therefore—also create a complexity that is difficult to grasp. Psychoanalysis has gathered a great many reflections that need re-thinking. This is possible in so far as the system of reference is clear and there are no important confusions in case descriptions between, for instance, drives and affects on the one hand, and libido/love and aggression/hostility on the other, to name only the most obvious examples.

Affect is both a mobilizing biological system of physiological energy and, at the same time, a signal system and thereby a system of *communication*. One of its functions, amongst others, is to regulate distance in social life (in relationships). An encounter prior to a sexual interaction is, in the animal world and *a fortiori* in the human world, a *complicated* process. Instinctual wishes ["*Triebwunsch*"], the choice of object based on visual, auditory, tactile, and other perceptions and on interactions having echoes especially in the past, as a reinforcement or as a process of reducing the mutual inductions of affects— all these play a complex part in it. These affective details should be analysed and not eliminated from the system of theoretical reference. Similarly, in seduction (a complex affective interaction involving several factors), it is also important to remember the parts played not only by joy and surprise, but by the *inhibition* of antagonistic interfering affects such as distress.

All these instinctual processes aim at satisfaction (pleasure principle) and tend towards satisfaction (e.g. in the sense of rediscovering the object necessary for the accomplishment of these processes, and thus for obtaining satisfaction). It is the supreme regulatory principle in this area.

The "display rules" of affects (Ekman & Friesen, 1969) owe a great deal to education, to the environment. In alexithymia (the inability to express affects), affective expression ranges all the way from intensification and demonstration of suffering to disintensification, to the total disappearance of the affect (e.g. of anxiety). The control of spontaneity may be introjected to a greater or lesser degree, so that the individual gains an autonomy over regulations compared with the external law that preceded it (Ekman, Levenson, & Friesen, 1983). This also corresponds to the direction taken by historical development in our culture (Elias, 1969).

In psychoanalysis, the subject learns, relearns, and, above all, becomes *aware* of affective communication in its capacity both to transmit and to receive. He rediscovers and will therefore consciously know his affects, especially in the original and fundamental ["*ur-sprünglich*"] context of his earliest fears and desires. This return to the fundamental, to the original, which is the furthest from consciousness, is traditionally conceptualized as "regressive". If by this is meant a step-back ["*re-gredi*" in Latin], a re-turn, crossing a space in the sense of going back, this "reversal" is in fact vital for *finding oneself* again and should not be compared pejoratively to the progressive, as though regression should necessarily have a negative connotation.

Studies have shown that the communication between mother and baby takes place through prosody, music, variations in tone, exaggerating the silences (Fernald, 1989). The psychoanalytic situation permits *reunion* with this kind of communication and, through it, a reunion with the *affects*.

Psychoanalysis can also be defined as the science of *symbolic forms* that *serve communication*—expression and exchange—of affectivity and the forces underlying it. It studies the specific *cognitive* constellations serving the expression of that affectivity (e.g. the Oedipus complex) and the forms of symbolism that, condensed, and displaced, transmit that *affectivity*, and then seizes it through interpretations. If cognitive forms belong to this science, it is always in a certain *perspective*, from the viewpoint of *affectivity*.

Visceral experiences and metaphors

To summarize, the *emotions* are well-defined, stereotyped psychophysiological and neuro-endocrinological states; they are unitary and transcultural, often deep-rooted, representing *subjective* states and, especially, a motivational pressure. They are also characterized by their *hedonistic* quality—that is, their subordination to the pleasure and unpleasure principles. They are expressed through different *channels*, either verbal or nonverbal; verbal does not necessarily imply intentional, any more than non-verbal implies non-intentional. The humours are probably defined by a multiplicity of emotions. Thus, for instance, in the depressive humour, or depressive affect, several emotions are found, such as sadness, scorn, some aggressivity, indeed hatred. The emotions are also *bodily*, visceral (e.g. the heart-beat), and motor (e.g. muscular tension of the voluntary motor functions) *experiences*. There is a whole imagery of autonomic experience (in the sense of the autonomous or vegetative nervous system). Its easiest expression is through the mediation of *metaphors*, hence the great interest in them for psychoanalysis. While apparently talking about one thing, the metaphor is often referring to the emotional state of the subject.

Emotion is at one and the same time a *state* and a means of *communication* through the expressive behaviour that governs it. There are probably a certain number of more *basic* emotions around which, in the semantic field, *nuances* are grouped. An emotion is both a genuine *bio-psychological* affective *programme* and a communication process involving different signalling systems (verbal and non-verbal, as I have already said). For many authors, anxiety and depression are states highly saturated with different emotions. I would be inclined to think that what we describe as clinical anxiety and depression are in fact composed of various emotions (over and above the anxious and depressive affects, e.g. contempt, disgust, etc.). As signal-affect, the anxious affect was, rightly, put in the forefront in Freud's system. Nevertheless, depression, too, has a *signal* function (demanding the withdrawal of the frustrating or traumatic situation) like all the *other affects*, joy for example. Dahl, Teller, Moss, and Trujillo (1978) propose a classification distinguishing between those emotions that refer to the "*object*", expressing attraction (love, surprise) or repulsion (fear), and those referring to the *self*, with positive expressions of contentment or joy and negative expressions such as depression or anxiety. However, at the present time every classification presents some difficulty, and we are far from pronouncing the last word on the subject.

Finally, let me emphasize that the communicative value of emotions is clear in their *identificatory* capacity—with the Other's joy for instance; that this is the child's mode of communication with his mother. It is these identifications that play an important part in all communicative processes and are fundamental in psychoanalytic communication, as a re-occupation ["*re-gredio*": regression] of poorly acquired positions in earlier affective communications, opening the way to understanding and interpreting them. On this basis a psychoanalytic theory seems feasible.

Defences

If it is considered that "defence" functions contrary to the affective mobilizing effect of information contained in an external or internal perception (mobilized drive and affect), for example against the sudden appearance of elements stored in the memory, the implication is that it always operates to defend against an affect with a negative connotation. This is a contemporary widening of Freud's first conception, which permits its integration into the information-processing model. In Freud's original conception, the defence operates against the sudden appearance of an instinctual drive. This view fits into a conceptualization characterized by the centrality of the scientific metaphor of the instinctual *force* operating within the subject. The new idea defended here takes into account defences against both internal perceptions, e.g. mobilized instinctual drives and external zones ("disavowal"—Freud, 1923e, p. 143), and against the sudden appearance of mnemic elements, as in repression. In Freudian works, there is a continuity from the first formulations about repression on the model of the multiple personalities of hysterics to the classical repression of Weinberger (1989) and to the hypercontrol in "alexithymics" described by post-Freudians (Nemiah & Sifneos, 1970)—almost going back to operative thinking (Marty & de M'Uzan, 1963).

The notion of *repression* already existed in the psychological tradition that would have been familiar to Freud, notably in Herbart (1816, 1824). The metaphor of censure (Freud, letter to Fliess, 22 December 1897) recalls the socio-political context of the monarchy (Haynal, 1983b). There are two discernible forms of defences. The former, whether they are substitutes for repression or rationalization, intellectualization or elimination by "explaining away" and "analysing away", are regulatory mechanisms operating within the subject, with the aim of reinforcing what is repressed and rendering it subjectively credible. The latter are fundamental mechanisms of introjection and projection, and they operate in the relationship with the other. The action of dismissing unpleasant contents from awareness may be *reinforced* by the intellectual control represented by *rationalization* or obsessional doubt, or by the reinforcement of opposites, as in *reaction formations*. The contents that have been pushed away may be *projected* into the outside world in the object relationship, or, as is the case of splitting, be alternatively *present* or absent from consciousness and so, at times, cut off from undesirable contexts. They may also persist in symbolic disguise in the various forms of sublimation. [Traditionally, repression is represented vertically towards the bottom, as opposed to splitting, which would be situated on a horizontal level. However, this spatial metaphor is not presented by Freud very consistently—he talks about "suppression" (1900a, p. 582; 1923b, pp. 23–24). Additionally, Freud's original text and its main translations present a variety of connotations: "*Verdrängung*" suggests the displacement of water by a boat, "*le refoulement*" has the image of pushing back an entering crowd, while the English "repression" has strong social, indeed political, associations.]

The aim of defence mechanisms is to avoid the displeasure of having to live with unpleasant thoughts, or to avoid a self-image that seems undesirable. In some way, therefore, they are related to the pleasure principle. This aim is achieved by avoidance in relationships and by distortion or exclusion of information within the self (e.g. repression, splitting, forgetting, rationalization, attempts at losing information or twisting it). These phenomena have, in part, been studied in contemporary psychology under the heading of "cognitive dissonance", "intentional forgetting", or "defences against perceptions". It is known (Pennybacker, 1988) that certain defences against recall have an effect on the immune system and make the subject's *health* more fragile. This, of course, provides an opening to psychoanalysis for the scientific understanding of problems of health and illness. Their study has highlighted the effective and structuring quality of defence mechanisms in "coping" processes and attempts at mastery. Others have to pay a high price in terms of communication with the environment; in that sense such people suffer from "maladjustment". The interplay of affectivity and the defences against it is a fundamental dimension for the psychoanalytic conception of human activity and one that will undoubtedly converge with research from widely different scientific traditions. Reflecting upon our own psychoanalytic discourse, particularly with reference to this subject, could, in the next few years, encounter some surprising resonances with a view to achieving a better understanding of human functioning.

CHAPTER FIVE

Text–context interpretation

Anything can mean anything . . .

Lewis Carroll, *Alice in Wonderland*

L iving is comprehending: if comprehending means receiving information from his surroundings, every living
being—whether equipped with a central nervous system
or not—receiving information that enables him to survive,
"com-prehends". One way of *apprehending* [*"ad-prendere"*] is to
put the information, the *"text"*, into the *context* in which this
text, this information, arrives. Whoever re-places, *interprets*,
and so introduces some comprehension on the receiver's horizon at that precise moment. He can clarify the text, introducing
new perspectives into it. In this sense, no interpretation can be
"complete". Learning ("cognition") is not an accumulation of
representations from the environment, it is a continuous process of transformations of behaviour and internal representations aimed at "living with" the environment according to
one's own internal representative ideals. Bion's (1962) "K" (for
"knowledge") is in effect a basic tendency of the human being.

71

Prejudice is rigid—creative interpretation is exactly the opposite. It is the combinative play of these representations that enables creative results to resolve problems in a *new* perspective.

If interpretation is an expansion of the discourse from what has been said, taking into account both the earlier context—and thus learning—and the emotional context of the communicative situation (in this case, the analytic situation), it is different from an explanation in a causal schema; it is, rather, the presentation of what has been said in a wider context—namely, in the personal history of the individual and the interactional situation. Interpreting is obviously only possible in a precise perspective—what Bion called a "vertex" (1965, pp. 103 et seq.) —a search for the unconscious nodes where the analyst feels anxiety or depression or the defences against them, such as inhibition and negation.

The problem concerns signifying information, in other words, meaning: how the meaning of a discourse full of subjectivity can be *apprehended*, formulated, "reconstructed", *expressed*, and put back into the discourse.

According to all the evidence, the coding/decoding model is of no help even if Freud also used just such an image in his telephone metaphor (Freud, 1912e, pp. 115–116), the importance of which for psychoanalytic *listening* cannot be overestimated. In fact, Freud explains that the analyst, with a sensitivity akin to the telephone receiver, catches the waves and makes sense of them through his earlier analogous experiences. Through introspection, he understands something of the Other—which is, moreover, at the root of his motivation to listen (Freud, 1950a, letter to Fliess, 14 November 1897). An understanding of the Other is possible in so far as one can identify with him, "put oneself in the shoes or skin of the other" —metaphors from everyday language, which well express these processes.

However, on the other hand, a more detailed examination should throw more light on these ideas of affective resonance and on the *analogical* thinking that is set in motion by "associations", in the relaxed state of "free associations" or "free-floating attention", which eliminates the tendency to dismiss from perception any nascent thoughts. (Analogical thinking, and particularly the relaxed freedom to receive it, seem to me to be the fundamental axes of free association and free-floating attention.)

We should, therefore, go beyond the decoding model, to ideas borrowed from *more advanced* linguistics, especially the idea of a contextuality permitting more pertinence and the notion of interactional models with "locutory [descriptive], illocutory [declarative], and perlocutory [soliciting action])" dimensions, as the theories of "speech act" from Pierce to Searles.

Reconstructing meanings—constructing a world

We interpret all day long. I see a machine to which I am unaccustomed but on which it says: "ON/OFF"; I interpret in the context of the machine that "ON" is the instruction for starting it. If I knew what the machine were used for, I could, *in the given context*, anticipate the result of pressing the appropriate button.

I interpret gesture, my neighbour's smile; I interpret the cross expressions of my rival. I "interpret", *in the strict sense* this time, the bored air of my analysand after several sessions following *another* interpretation I had made to him that he did not like. I thought that by interpreting I could centre our attention on the problem that had blocked him, which, as a result and through defence, made him put on the bored expression that irritates me.

I interpret my irritation, in turn, plus the image evoked by my irritation, of a scene from my childhood. At that time, Zoli had spoiled the pleasure I had in playing. Why was I thinking just then about that little redhead whom I have not remembered—at least "consciously", by which I mean not having given it any thought—for decades? A young friend who smelled bad. . . . At this point, the "hermeneutic" circle closes, only to re-open immediately: I understand how I defend against all these negative feelings, thus participating in my analysand's sulkiness during several sessions. A sulkiness that I recognize, we must both sort out. "Both" brings me back to my analysand, and the spiral goes on, the game goes on, the analysis unterminated, endless, like life. Each act of comprehension is also an act of interpretation: I see myself on my death bed; a doctor, leaning over me, whispers something to the nurse—yes, she is present—and I understand that the end is not far off. Looking at my doctor's face and searching for meaning in it, I try to guess. . . . I interpret the scene, I interpret to the end. . . .

Pausing there in my rêverie, I reflect, I interpret: in the end it is a history of a "negative" smell, the smell of death.

But how many other configurations of smell do we know—erotic, seductive? We can name them; sometimes it is *necessary* to name them. Psychoanalytic communication is a verbal and conscious communication, as far as is possible. "Where Id was, there shall Ego be"—through conscious thought in the auditory sphere. But let us not forget that the "Id was". The thousand and one non-verbal ways of communicating must first be there in order to be *translated* into verbal language, interpreted so that consciousness may thereby be enriched.

Interpretation is the translation from one language to another. A piece of music is also interpreted. In other words—as Freud's mother-tongue so well expresses it—it is a *transposition* [*"Übersetzung"*]. This also implies that what is transposed is not more true, more worthwhile than, or superior to the original, it is not completely different, not entirely transformed. It is simply put into another context, transmitted, formulated by another person who talks from another place.

Transposition has its *limits*. Dan Sperber (1975) emphasizes that symbolism is not the equivalent of what can be transposed in conceptual codes, it goes beyond it. It is not just a matter of being able to translate in the discourse what can be translated into other symbolic representations, but also, quite simply, of recognizing its relevance, its validity, and of assigning a place to it in the memory (p. 113).

Interpretation manifestly is not an exclusively linguistic problem. It is perceiving and "decoding"—understanding and giving meaning to the communication, both verbal and non-verbal. There is a transmission of emotions, which, obviously, is not linguistic, the communication of a text, which must also be understood in the *context*, for example of the "scenario", and in the diachronic perspective of that context. The sciences that study communication, from rhetoric and linguistics to those concerned with narrative and artificial intelligence, can offer clarifications to the analyst's work, especially to his interpretative work.

These sciences—and modern philosophy—redeem us from a naive conception of "reality" according to which the internal world would be the image of the "real world", as Bertrand Russell and the young Wittgenstein would still have been teaching at the beginning of this century. The mental *construction* of "his" own reality is made in a precise context and by means of his earlier knowledge. In the seething Vienna of the nineteenth century, Herbart, one of the *Masters* of Freud, noted that every perception is only "apperception" [*"ad"* + "perception", a perception *added* to another element (of the memory) already present: Herbart, 1816, 1824]. The "reality" of the external world—and *a fortiori* of the internal world—is constructed in this way by the constraints, limitations, and influences of the psychic continuity and *interaction*, indeed the *language* of human beings.

If Edward Sapir (1929) is right in saying that language is one of the most important determinants of our way of thinking, then changing our expressions, our language, is also a start in changing the possibilities that we have of understanding ourselves and of being understood, and thus of our way of being with ourselves and with others.

> Human beings do not live in the object world alone, nor alone in the world of social activity as ordinarily understood, but are very much at the *mercy of the particular language* which has become the medium of expression for their society. It is quite an illusion to imagine that one adjusts to reality essentially without the use of language and that language is merely an incidental means of solving specific problems of communication or reflection. The fact of the matter is that the "real world" is to a large extent *unconsciously* built up on the language habits of the group. . . . We see and hear and otherwise *experience* very largely as we do because the language habits of community *predispose* certain choices of interpretation. [Sapir, 1921, p. 75]

The first problem is of *communication* (including the subject's communication with himself), then the *understanding* of it through another subjectivity. One of the key ideas in this field is *context*. Douglas Hofstädter (1987, p. 44) could write that the conception—especially George Bowl's and that of the majority of researchers on artificial intelligence—that ". . . the laws of thinking 'consist of formal rules for the manipulation of propositions'" has been shown to be wrong. It seems that it is in the understanding of a *context* that a communication becomes pertinent. The semiotic "coding/decoding" model should be completed by that of "inference" operating from the *context* (Sperber & Wilson, 1986).

One of the criteria defining the validity of an interpretation is, obviously, the *context*. But there can be several obvious interpretations, and I do not think that there is one single correct interpretation for one particular situation. Interpretation introduces a *structuring* thought and steers the dynamics of the interactions in one particular direction, to the exclusion of others. The phenomena of "multiple appeal" (Waelder, 1930) and "multiple function" (Hartmann, 1947, p. 41) are such that, even if the model of the psychoanalyst, the supervisor, or the reader of a written account is different, it could well be that the same process is used throughout various models.

The Greek word for "interpretation" is "*hermeneia*". "Hermeneutics" is thus the synonym for "interpretation". If Freud said that psychoanalysis is, in the first place, the art of interpreting, a "*Deutungskunst*" (Freud, 1923a, p. 239), it could be deduced that one does not have to make an exegesis in order to learn the truth from untouchable texts, but, rather, to be a "mediator". The Greek word itself, according to Gusdorf (1988), refers back to the god Hermes, the *messenger* between the immortal gods and the human beings, patron of communication, and, as it were, symbol of the *circulation of meaning*.

Gadamer (1975) writes that "*eikos*", verisimilitude, likelihood, enlightenment ['*das Einleuchtende*'] are all part of a series that defends its legitimacy: and, according to Rorty (1980), the practice of hermeneutics is more like "getting acquainted with a person" as opposed to following a logically constructed demonstration (pp. 318–319). If interpretation is *reconstructing significations* from more or less crude or complex traces, the concept of "inference" from modern logic clarifies this process well. The analysand's words are taken seriously, as if they were a "sacred" text, that text being the manifestation of a meaning.

The interpreter tries to steal this meaning, to know what is *implied* in the communication that has been made.

Interpretation is in part an "inference" from the logical point of view—a non-deductive form, as I have just been saying. The validity of such "inferences" is no longer contested in contemporary logic (Genesereth & Nilsson, 1987). Making an interpretation appear in this perspective opens up a possible area of interdisciplinary collaboration. (Such collaboration has not yet been achieved, almost certainly because psychoanalysis is rather poorly represented at scientific meetings. Its representatives are often "sectarian", in that they show an aggressive assurance, which goes hand in hand with a refusal to submit their presuppositions to methods of verification, or at least of assessment.]

The hypothesis that we copy the contours of the world within ourselves seems to me largely outmoded. In actual fact, men *construct* their world. Our image of the external world, and of our interpersonal relationships in particular, is a complex one, made up of projections, identifications, and especially projective identifications, and it takes a complex model to describe such complexity adequately. Psychology has shown quite clearly that our perception of the external world has already been tainted by our own subjectivity, by our expectations and projections.

Furthermore, in talking about our own experiences, the language, of necessity, becomes metaphorical, and the concept of *metaphor* is accordingly fundamental to the understanding of the discourse about the internal world. *Every* language uses *tropes*. Their relative importance is controversial, but the impressive increase of literature on metaphors is proof, as far as researchers are concerned, of its impact on the same problematic areas in which we are interested. Metaphors give birth to "inferences". If metaphor is used in speech, it implies the usage of comparison, with all the attendant risks of misunderstanding. But metaphor also suggests form, Gestalt—contours, overtures—necessary to the formulation of an *experience* and the ability to communicate it. This experience and the attempt to formulate it are precisely the subject matter of psychoanalysis.

Since knowledge can be either theoretical or an elaboration of experience, it must be made absolutely clear that psychoanalysis aims only at knowledge *after* and *about* an experience. This distinction between the two kinds of knowledge—theoretical and experimental—has already been advanced by Bertrand Russell (1929), who talks about "knowledge by acquaintance and knowledge by description" (p. 209). In the latter, we are acquainted with something simply by description; in the former, we have an experience of it, we have "frequented" it. Ferenczi made a similar distinction in the problem of analytic treatment, and he declared that true knowledge must be based on the *experience* of the subject's internal world and the relationships that he is able to make with those around him in his environment.

It must be remembered that metaphor is one of the *constituent* principles of language (Searle, 1979a, pp. 92–93). When a patient tells me he adores gadgets that can shoot up the aerial from inside the car and, as explanation, adds that he also likes conjurors who spit fire, it can be assumed that he is trying to transmit affectively through these metaphors his pleasure at something that appears suddenly and has the shape of an antenna or a tongue of flame. In effect, he makes me feel, through the metaphor, the background of what he says, which at that point is still unconscious for him. Only the analytic work will enable this passive boy, full of avoidance mechanisms, to take account of that pleasure and the meaning of the metaphor.

Searle (1979b) makes a distinction between the "speaker's utterance meaning" and the "word or sentence meaning". A third should be added: *implications* that may, at the time of their formulation, not be conscious for the subject. Psychoanalytic interpretation aims, in the first place, at these "implications" read in the context (of precedents and non-verbal communications).

The phenomenological method—an approach that helps understand man and his behaviour in the perspective of his actual experiences—is the child of the twentieth century. Before that, such a method had not been developed. But it should not be forgotten either that in this art of interpreting, as Freud qualifies psychoanalysis on many occasions, in this phenomenology, he *introduced* the idea of *forces* that provide a system of references for understanding man, his drives, his constitution, his defences, and the implications of his affective experience. Although, historically, psychoanalysis was not regarded as hermeneutics, it does constitute a particular variant of it. Romantic hermeneutics was part of the ambience of Vienna: the understanding of others in general, understanding as reactualization, interpretation, the "donor of facts", the dynamism of meaning, are all themes that were important to Schleiermacher and that became, in another guise, part of psychoanalytic thinking. To my mind, though, psychoanalysis does use the hermeneutic method, as we shall see, to create a scientific discourse, which makes it *more* than pure hermeneutics.

According to Rorty (1980) "hermeneutics is the study of an abnormal discourse from the point of view of some normal discourse—an attempt to make sense of what is going on at a stage where we are still too unsure about it to describe it . . ." (pp. 320–321—Rorty's "abnormal discourse" refers to what does not deliver up its meaning immediately: a mysterious text).

I am close to thinking that we can examine it by the Socratic method, and that in doing so with interrogative interpretations, we establish connections (Rorty, 1980, p. 360). Hermeneutics thus becomes the study of the "unfamiliar" (p. 353), in order to make it familiar, comprehensible.

Before Schleiermacher, hermeneutics was the science of texts; with him, it became an attempt by a member of a given culture to *understand the experience of another*. In actual fact, each of the protagonists in psychoanalysis, enclosed in his or her own system of references, finds difficulty in understanding the more intimate depths and references of the other. So when we talk about hermeneutics in psychoanalysis, we are, of necessity, also talking about *intimacy* with another, empathic understanding, being in tune [*Einfühlung*], which in turn is expressed in the analyst's words. Interpretations cannot, therefore, be separated from empathic identification, from countertransference, nor can hermeneutics be dissociated from the *emotional experience* surrounding the words.

It is fashionable to emphasize the structuring aspects of the "narrative", of the construction of the *narrative*, which is a way of reconstructing a self-image, a self-representation, a schema, a self-coherence, and relations with others in the course of psychoanalysis. The subject which will be inhabited at the end of the analysis. . . . Yet the advantage of the narrative is its *ambiguity*. The dream narrative is full of "lacunas", which provide a chance to "work" with them (such work is called "elaboration", "working-through"). This helps to construct a narrative that undoubtedly contains some historic truth but, like a fable, permits the expression of how one feels, how one would like to feel, to be. Its ambiguity is its strength. The archaeological metaphor refers back to preexistent information, which surrounds the precious fragment snatched from the past, and which tries to integrate it in the fabric that, in our case, the analyst and analysand are creating together; it is what leads the analysand to see himself in a new light.

Countertransference, in this context, is the analyst's instrument of perception and affective echo, a precious instrument for understanding the Other. Neyraut (1974) said that countertransference *precedes* transference; in the same way, the analyst's feelings and needs precede the creation of his professional surroundings and the psychoanalytic situation, all of which must be taken into consideration in the analytic setting that will be created. It seems pointless to wonder whether or not the term is rightly chosen (Brenner, e.g. 1984). What *is* fundamental, as Balint said, is the analyst's *contribution* to the *creation* of that situation and, as has also been said, not simply his understanding of it, and, above all, that this understanding passes through his internal instruments. The psychoanalyst has his own "narrative" structure: it is his theory (either Freud's Oedipus myth or Melanie Klein's child battling against the death instinct). The analysand also has his own narratives, however inconsistent they may be, presented through timid suggestions. From the general narrative of the analyst (his orientation) and the incomplete and provisional one of the analysand, the work of analysis is the building of a *new* narrative structure.

In this context, the notion of *schema* acquires importance. It is a notion that derives from Piaget as well as from contributions to cognitive psychology and to information processing theory. It can be considered as a refinement of the *representation theory*. The schema is a unity of already constituted cognition in which the subject assimilates—or attempts to assimilate—new elements. Obviously, such a move enables some *saving* of effort in the cognitive *work*. In the field of what is known, the schema makes perception easier and faster. Its dangers—of assimilating the unknown into a schema already known, without question—are obvious. The whole psychology of prejudices, of schemas made of enemy and friend (which I examined in the context of fanaticism—Haynal, Molnar, & de Puymège, 1980), is illustrative of these dangers. The underlying cognitive mechanisms can profitably be studied from this point of view. Indeed, transference, the confrontation with fantasies, often makes it possible to detect some of these mechanisms and, in uncovering them, to gain freedom from the rapid "schematizing" supported by often powerful affects. The example quoted shows, at the same time, the advantage of the psychoanalytic procedure over many other techniques: a cognitive technique alone would only interpret the mechanism, certain existentialist approaches only the affectivity. Freudian scientific traditions take account of *both* the error in cognition and the underlying *affective* economy.

* * *

Translating what the analysand communicates so that it can become what he thinks is called interpretation. Interpretation questions; it does not put a closure on a subject, nor is it there to *assign* truths, especially truths that would then be more true than the analysand's own truth. It is his truth that must be explored.

It is undeniable that theory influences interpretation, as does all the analyst's previous experience, including his reading, his thinking, his personality. However, when it is no longer a thought that arises in a creative movement but, rather, the veneer of a *cliché*—when, in other words, his alleged preexistent knowledge takes the upper hand—then interpretation is no longer what it should be. Remember, this was the criticism Raymond de Saussure levelled retrospectively at Freud in 1956: that he was "too preoccupied" with his theory; this should become a lesson to us. Michael Balint made the same comment about his analysis with Sachs ("My analysis? I was very dissatisfied with it . . . it was rather theoretical"—interview with Bluma Swerdloff in August 1965, Balint Archives, Geneva).

Such an interpretation can only be given by the "unobtrusive analyst" (Balint, 1968b). So, interpretation rests on the analyst's lips as the expression of an idea or a question about a possible meaning, and not an authoritarian act, telling the other what he thought. Comments of the "You think that I . . ." sort can become very intrusive, especially if repeated.

In 1958, Loewenstein, a sensitive analyst with a great interest in psychoanalytic practice, was already writing: "I doubt whether anyone has ever carried an analysis through to a successful end without having done anything else but interpreting", and this was at a time when the idea of interpretation as the instrument of the "real" was at its height and often being proffered as immutable dogma. In reality, the analyst does much more; he creates an environment, provides "holding", reflects echoes, and communicates much more complex signs than one would care to admit in a simplistic theory. I have no doubt whatsoever that interpretation is a very important tool, leading to an awareness of painful contents and the defences against them, and that it can also mark the entry of affects into the kingdom of language. But to say, as Loewenstein does, that interpretation alone is not psychoanalysis can pass as a corollary of what has been said about its importance. Transference is one of the mainsprings of analysis, in so far as it is one of the principal elements in recognizing the *meaning* of scenarios, unconsciously recreated and communicated in the analytic scene. In a more advanced phase of the analysis, the patient will make his own interpretations, more and more independently, thus anticipating the end of "being in analysis" and embarking on another stage of endless analysis—namely, self-analysis.

The transition to science

One of the methodological problems of Freudian psychoanalysis is undoubtedly that of generalizations. Freud wrote in 1897: "I have discovered in myself a feeling of being in love ["*Verliebtheit*"] towards my mother and of jealousy towards my father" (Freud, 1950a [1887–1902], letter to Fliess, 15 October 1897). Later on, taking into account mythological material and material drawn from patients, he *generalizes* and adds: "I think this must be a general event [*"allgemeines Ereignis"*] of childhood" (ibid.). Yet, even if he found the same configuration in himself through introspection as he found in some patients, in children, or in others, and in the material left behind by some of humanity's greatest representatives (such as Sophocles' Oedipus and Shakespeare's Hamlet), one might still query the importance of this "complex" constellation, its *centrality*, and its generalization. Freud did not fall prey to this; if he supported the ethno-psychoanalytic researches of Géza Róheim—financed, incidentally, at his instigation, by Marie Bonaparte—it was to satisfy himself of the legitimacy of those generalizations. Otto Rank's studies on mythological themes were of a similar order. But neither Freud nor especially his pupils always proceeded with such caution, and insights from *personal experiences*, which often seemed to have a great *impact* with some of his patients, were rapidly raised to the level of generalities with limitless validity. Moreover, this process of generalization has throughout its history been characteristic of the use of psychoanalytic concepts. At first, they may well clarify one *aspect* of the psychic life of *some* subjects; but they rapidly become a generalized law postulated for all patients in all circumstances of life. So, mechanisms that once were proposed as being characteristic of psychotics, or others as characteristic of so-called borderline patients, became concepts considered as highly important for *all* analysands. Gradually, the links between these concepts become wider and wider: the net, which at first caught only specific fish of a certain size, gradually holds more and more phenomena (such as "narcissistic traits" or projective identification as all-purpose explanations). The situation is undoubtedly a reflection of the psychic complexity of the human being, but it also points to a major *methodological* problem of the passage from interpretation of oneself or another (and if of another, then *through* oneself) to the *construction of a science* based on such interpretations. There must, first of all, be a hermeneutic network allowing some degree of understanding, and only then, with great care, can hypotheses be formulated, in the hope that they will become "laws", fundamental aspects of psychic life. In order to make this leap, the most rigorous scientific methodology must be applied: data must be accumulated, screened, linked, and correlated; comparisons must be drawn between different methodologies with respect to convergences and divergences; and a scientific model that is capable of such complexities must be developed. From this point of view it seems to me that psychoanalysis is only at its beginnings. Developmentally, science—especially artificial intelligence—has only very recently managed to *offer a true scientific examination of such complexities*. This is also the case for linguistics, ethology, and other sciences that are no further advanced than psychoanalysis in this respect.

If, despite the understanding of human phenomena through the narrative and its interpretation, the *leap* to the construction of a science has not taken place, if the "Project for a Scientific Psychology" (1950a [1895]) did not see the light of day during Freud's lifetime, it is because this longed-for *transition to science* proved to be premature. Chapter 7 of *The Interpretation of Dreams* (1900a), some commentaries on drive organization in the *Three Essays on the Theory of Sexuality* (1905d), and some other scientific contributions by Freud are isolated fragments, like stones in a post-modernist house. Without doubt, his general understanding has, intrinsically, a very great *narrative* and *interpretative* value, but basing a scientific construction on it proves more difficult than was apparent in the enthusiastic beginnings.

Describing one of the major results of the psychoanalytic interpretative method, Freud (1905e [1901], p. 15) wrote: "I have thus learned to translate the language of dreams into the direct and habitual mode of expression of our thinking". Describing one of the major results of the interpretative psychoanalytic method, he expresses the hope that psychoanalysis, like archaeology, will bring to light ". . . after a long period of concealment, the most important traces, albeit distorted, of antiquity", and he adds soberly: ". . . I have completed that which was incomplete but, like a good archaeologist, I have not neglected to distinguish in each case that which I have added from the authentic material." A description of incomplete endeavours was bequeathed to subsequent review with intellectual tools that had not yet been developed.

From the intimate to the scientific

It is worth remembering that Freud's methodological descriptions are sometimes based on positivistic observations (for example, his description of the stages that lead to the construction of a theory in "Instincts and their Vicissitudes", 1915c, p. 117), but that in other passages he relies on ideas, even general presuppositions, such as what he calls "the witch Metapsychology" (Freud, 1937c, p. 225). In short, he has a range of approaches both *inductive* and *deductive*, positivistic and rationalistic. Since he lacked the epistemological tools, in matters of scientific methodology Freud was at his time, perhaps despite himself, a more intuitive than explicit innovator. He sometimes expressed the hope that his constructions would find an unexpected confirmation in later discoveries in the biological sciences, as in this passage written with Breuer (Freud, 1895d): "In what follows little mention will be made of the brain and none whatever of molecules. Psychical processes will be dealt with in the language of psychology; and, indeed, it cannot possibly be otherwise" (p. 185). At about the same time, in 1896 (30 June) he wrote to Fliess: "Perhaps I have found . . . the ground on which I can cease my psychological explanations and begin to rely upon physiology" (Freud, 1950a; see also Freud, 1914c, p. 78—although he was also opposed to the idea of the subordination of psychology to biology—see, for example, his letter to Jung, 30 November 1911). This idea of *convergence* has since been taken up by post-Freudians (for example, Bowlby, 1981). Today, the possibility of a model based on "artificial intelligence" offers the opportunity to test the coherence of a model and to open up new perspectives. Claude Le Guen (in Le Guen, Flournoy, Stengers, & Guillaumin, 1989, p. 7) maintains that psychoanalysis is a science. Olivier Flournoy, however (in Le Guen et al., 1989), takes a contrary view:

> In my writing, my aim is to describe the secret, intimate face of psychoanalysis, the one I encounter every day with my patients, and which seems to me incompatible with the scientific approach, which distorts the very foundation of it, that is to say the unconscious. Claude Le Guen turns to the external world arguing that psychoanalysis must be a sci-

ence in order to take its place in the contemporary world. To my mind, far from being exclusive, the two perspectives should be able to complement each other. [p. 45]

My own concern bears on the problem of knowing how one moves from the "secret and intimate aspect" to the opposite position, in which psychoanalysis "must be a science in order to take its place in the contemporary world". It is with this *transition* that I am dealing here.

The methodological uncertainties in Freud's work and their poor reception (partly, in retrospect, exaggerated—see Kiell, 1988) contributed to some extent to the fact that Freud himself, as well as his followers, took a *defensive* attitude towards those who questioned the scientific "proofs" of psychoanalysis. The *rupture* with Academia, the fact that psychoanalysis developed in relatively *closed* circles and was organized, historically, as a "*movement*", in the manner of other "isms" of the time—politically: communism, socialism, Zionism; artistically: expressionism, surrealism; in the religious field: Catholicism, etc. (see also chapter ten)—may have facilitated the *deepening* of clinical work, but the difficulty of providing proof was almost short-circuited. The replacement of proof by the consensus of a majority of initiates already convinced—especially in local groups—with the development of *local orthodoxies* provoked particular phenomena. So, passionate disputes broke out—for example, between Vienna and London, and the two protagonists, Anna Freud and Melanie Klein, over the issue of child analysis. The affective charge between them is understandable when both were seeking a truth supported by peer or sometimes paternal *recognition* within the group. Every "deviation" seemed a threat to that truth upon which high expectations and personal happiness depended.

The search for clarifications: structural anthropology

Genuine exchanges between representatives of *new approaches* in the human sciences, in epistemology, and psychoanalysis (such as have been realized, for example, between scientists like René Thom and psychoanalysts—International Colloquium on "The Unconscious and Science", UNESCO Palace, Paris, 5 March 1988; Dorey et al., 1991) should achieve some clarification of the epistemological foundations of psychoanalysis on the one hand; on the other, particularly through the contribution of artificial intelligence, it should find coherent models opening the way to some interdisciplinary agreement. The fact that Freud usually refused to consider statistical methods as valuable in his field (Freud, 1916–17, p. 461; 1933a, p. 152) and that he often declared that one of his methods—other than clinical observation—was that of "speculation", no doubt hid an intuition: to avoid for psychoanalysis the sterile confrontation with the behaviourist laboratory and its inevitable excesses of "terrible simplification"—in other words, with a method unsuited to the complexity of the psychoanalytic encounter. [Freud's position seems to oscillate somewhat. Early on, in order to support his opinion on the etiologic role of heredity in the neuroses, he wrote explicitly: "Our opinion on the etiologic role of heredity in nervous diseases has to result from an impartial statistical examination and not from a 'petitio principii'" (Freud, 1896a, p. 143). Moreover, he demanded statistics from Rank to support his theory of birth trauma, and Jones (1957, Vol. 3) states that in his opinion this was "the only occasion on which Freud seemed favourable to statistics in relation to psychoanalysis". It is interesting to note in this context, that in the paper "On the Aetiology of Hysteria" (1896c, pp. 200–201) he makes an error in the statistics, speaking of 18 cases out of 20 as representing, in his words, 80% (instead of 90%). It could be deduced from this that at the time he did not attribute great importance to this data.]

An interesting *model* of interpretation for human sciences has been elaborated in anthropology, especially in Claude Lévi-Strauss's system (1958). (Ethnographic interpretation was ahead of psychoanalysis in clarifying its methodological and epistemological positions.) Using as example the Oedipus myth, Lévi-Strauss takes up the elements common to the different versions: the over- or under-estimation of blood ties, the theme of death through monsters, and difficulties in walking. Different circumstances surround these elements. In other words, for Lévi-Strauss, this myth, like others, expresses some of man's problems, such as his questions concerning the links within the family and sexual relationships. In his thought processes, his *attention* (his "listening") constructs his discourse on those *elements* that he judges appropriate for his aims (to retain the *universal traits* of the text). The psychoanalyst's attention, leading to another choice, also takes elements that seem to him, in relation to *his system of reference*, significant or specific for the individual to whose discourse he is listening. The analyst's system of reference is built, on the one hand, on his general experience and theory, and, on the other, on his understanding of important elements of the unique individual (the analysand). This process seems to emerge as fundamental to any theory of interpretation. In this way of proceeding, as in archaeology, the interpreter takes basic elements and puts them in a perspective that is his, thereby presenting a new meaning for reflection.

If anthropology is the science of the *otherness* of other cultures, psychoanalysis expresses the otherness of the individual, of the *Alter* in the same culture. Comprehension of this otherness can only be achieved with methods related to hermeneutics, even if psychoanalysis must obviously not be *reduced* to this. In fact, to approach a description of mental functioning, it can resort to "cross-validations", or reciprocal confirmations (for example, achieving a theoretical concept of the libido through observations of child development, etc.).

Freud (1933a) compares the id with chaos, "a saucepan full of boiling emotions", "uncontrolled passions" [*"Ein Kessel voll brodelnder Erregungen"* (p. 73); *"ungezähmter Leidenschaften"* (p. 76)].

At the same time, he considers (Freud, 1915e) that the unconscious confronts us with the "results of thoughts, the elaboration of which remains hidden" [*"Denkresultaten, deren Ausarbeitung uns verborgen geblieben ist"* (pp. 166–167)]. He tries to understand the untamed energies, chaos, the subterranean world of a *language*, of a different way of thinking, and to integrate it all according to what he understands as the "laws" of the unconscious. If the drives can be grasped through representations that they overdetermine (1915e, p. 176), the mobilization of affects is signified by cognitive constellations, and thus the understanding of the rational by the irrational becomes possible: "Where Id was, there shall Ego be" (1933a, p. 80). Otherwise, our knowledge is gained through indirect communication (non-verbal) or failures in verbal communication (slips of the tongue, parapraxes, dreams). Lévi-Strauss's project (1955, pp. 62–63) of seeking to translate without losing the sensitive in the rational is situated, in the end, in a continuation of the Freudian tradition.

Although Freud (1923a) thought that psychoanalysis was a "*Deutungskunst*"—an interpretative art (p. 239)—he also believed that through the practice of "*Deutung*", the analyst discovered certain laws of mental functioning and certain *structures*, and that he was thereby capable of constructing a nomothetic science of the human mind. Psychoanalysis is therefore not *only* a hermeneutic of text, or even of context, nor of verbal or non-verbal language. It is also founded on an *anthropology* of the Other through accumulated and formulated experience. In other words, it is a science.

With regard to the structural anthropology of Claude Lévi-Strauss, as with psychoanalysis, it has been argued that it is rather a matter of *conjectures* than knowledge based on clear connections between undeniable facts, an attempt "to understand the organizational principles underlying behaviour" (Tyler, 1969, p. 3). These criticisms bring us back to the necessity of presenting proof that validates these conjectures and "inferences", for example in the form of *predictions* (personal communication, P. Hermann, 1980). In this way the scientific status of psychoanalysis could be greatly consolidated.

Information processing

The introduction of the information model into psychology replaced models inspired by behaviourist and experimental psychology, all of which are, in the last resort, reducible to a stimulus/response model or, at the very most, a stimulus/organism/response model, and some elementary structures of perception. In other words, whereas one of the "assets" of behaviourism and its successors in scientific circles was *methodological* rigour, that rigour was obtained by simplification.

The new models, while remaining methodologically acceptable, were able to take account of complexities that had previously been unimaginable. Researchers in the past could only work with a limited series of data. The dream of each was: "If I had *all* the *data* at the start, I could make predictions about behaviour." This was possible with only rough facts in very limited areas. Today it is possible for the situation to change, thanks to models enabling processes to be examined, among them interpersonal ones: promising starts can be pointed out (Horowitz, 1988a).

Scientific psychology, under the influence of information technique, comes close to models able to study a complexity including strategy, aim, and memory. Furthermore, the emphasis on information has opened a whole dimension that is no longer included in the perspective of *energy* or *matter* displacement, but in the perspective of *information* per se. These views of information procedures may well be modelled more on a view of cognitive information displacement. But psychoanalytic theory, by its interest in *emotion* as well as *representation*, could contribute to a more balanced approach.

Information technique has brought to scientists and public alike the discovery of "*software*". It becomes an imaginable hypothesis that man, too, has a programme that is modifiable, at least in part, and not refused from the start in the name of static anatomico-histological or physiological "old-fashioned" science, where only the "hardware" is taken into account. In addition, information technique has provided a model for handling complexities: researchers engaged in strict methodology are able, through these new models, to approach complexities for which there has previously been no model (Singer, 1989).

Knowledge is organized in science by the creation of some "order", marking off facts, elements; in psychoanalysis, by elements common to several individuals: "We know that the first step towards attaining intellectual mastery of our environment is to discover generalizations, rules and laws which bring order into chaos" (Freud, 1937c, p. 228). Regrouping these facts in a *concept*—such as narcissism, for example—is a second step. A concept involves a statement declaring that in certain conditions, certain reactions appear with a certain probability. A simple linear causality will certainly not be able to account for the complexities of the human internal world, and, by the way, such causality has no longer any exclusive compulsory validity in modern science. The concepts of information processing and the epistemological models of the contemporary sciences—particularly physics—may be of help to psychoanalysis: it is obvious that only by taking account of the complexity of circular causalities, feedback, partial conditions, multi-causalities, will a model finally be attained from what psychoanalysts have been able through their discourse to learn about the human being.

A science has come into being that examines some part of the world on the basis of information processes. The idea of a system of symbols is at the heart of it. These *symbols* are modifiable—there are processes in the course of which symbols are found to have been created, modified, destroyed, and recreated daily. The most important characteristic of these symbols is that they designate objects, other symbols, or models. . . . When they designate processes, they can be *interpreted*: interpretation implies the realization or accomplishment of the designated process. This may furnish the psychoanalytic concept of interpretation with an *analogous* meaning: the anticipation of an awareness of what is implied in the symbols presented by the analysand which are the object of the interpretation (in this case, the analyst's).

This does not mean that one has to accept artificial intelligence unconditionally or embrace all its hypotheses or all its ambitions (such as creating an intelligence identical to the human one) in order to *take something from it* and to clarify in a new way phenomena that are still insufficiently grasped.

Parallel processing

When a child traps its finger, one can refer to his general vulnerability, his "programme" for protecting his bodily integrity, and even see in this particular subject a point common to all living creatures, at least those in the animal world. Yet, in the perspective of *parallel processing*, this perception brings about fears of another order, which psychoanalysis has called "castration anxiety". These fears may remain unconscious or enter the subject's consciousness and be assimilated in a current model. The parallel-processing model makes it possible to integrate the two paths, between other unconsciouses, for the first time since the beginning of the century, in a scientific vision that would probably be shared between several sciences through different modern disciplines.

Madame Singer cannot utter swear words in her mother tongue, but she is quite capable of doing so in French, her adopted language, which, moreover, she speaks perfectly: an illustration that the text cannot be dissociated from its context, especially its emotional context, or from the history of learning and the associations with which it is linked. At the same time, it is undeniable that in the analysis of notions such as representation, words and affects correspond to realities. (I am using the word "analysis" in all senses of the term: "to break down into elements", as well as the sense in which, historically, Freud used it for the mind.) [For the moment, I am leaving aside the question of the "nature" of reality: whether it is functional and, if so, to what degree—the problem of what is called "mentalism", the hypothesis of mental entities.] It can also be said that in the terms of information theory and from the viewpoint of parallel distribution processing, the aim of the different systems that are processed in parallel—systems such as affectivity, verbal expression, and action—is to function "in harmony". This is a reformulation of the thesis of mental conflict in another perspective, making it apparent that some harmony between affectivity, the text that is uttered, and action is the "aim" of those systems.

Although he did not invent the concept of *associations*, Freud did use them as a main instrument for understanding the mental organization of man (Rapaport, 1938) while being aware of the inadequacy of the model proposed. This is in contrast to the behaviourists, who, from Watson and the conditioned reflex to Skinner and reinforcement, remain faithful to pure associations. Freud has throughout constructed functional unities, interactions. He assumed "agglomerations" of associations in his topographical model, emotional and motivational forces that push certain associations to the fore in his economic model, and the historical continuity of those associations in his dynamic–ontogenetic model, leaving room for genetics in his idea of a "complementary series", under the rubric of constitution. This meets with the idea of "modules" in contemporary information theories—functional units in which recall by resemblance and parallel processes enables the construction of models for Freudian ideas. It is both impressive and promising to see how these systems and sub-systems correspond entirely to what the *avant garde* of present-day science advances as possible models of information organization, i.e. of our knowledge. Marvin Minsky, in his theory of a "society of mental entities" (1985), refers to functional organizations, which he calls "modules"; it is striking how close these are to Freud's topographical model of the mind.

Modules of the memory

Perhaps *communication* theory, studies on information processing, and the models of artificial intelligence can contribute to the metadiscourse of psychoanalysis in such a way as to make it more coherent and gradually enable its basic tenets to be clarified. In several areas there is surprising convergence among these studies, which, it must be remembered, are based on computational and cognitive functional models. For example, Fodor's hypothesis (1983) on the modular functioning of the memory is very near to Freud's "multiple inscriptions" (Freud, 1950a [1895], passim; then 1900a, pp. 537–542; 1915e; etc.).

The problem of traditional psychoanalytic interpretation is closely linked with that of *memories* and the nature of *memory*. In the current scientific conception of events, "Memory for whole events is stored widely, not in a single localization; . . . so that recollection of past events is a reconstruction from fragments, not a veridical playback of past events"! (Squire, 1987, p. 77). These words can be related to what Freud wrote in 1909: "We must above all bear in mind that people's 'childhood memories' are only consolidated at a later period, usually at the age of puberty and that this involves a complicated process of remodelling analogous in every way to the process by which a nation constructs legends about its early history" (1909d, p.

206). Similarly, "Piaget's abduction" shows the extent to which memory recall necessarily involves the memory being put into another context, which, if the case arises, even creates a meaning that it might not have had originally.

> For example, I myself have a very clear visual memory, detailed and vivid, of having been the object of an attempted abduction while still a baby in a pram. I can still see a series of precise details of where it took place, the struggle between my nurse and the abductor, the arrival of passers by and the police etc. However, when I was 15, the nurse wrote to my parents that the whole story had been made up by her and that the scratches on her forehead were her own doing etc. In other words, for five or six years, I must have heard the story of that abduction which my parents believed, and drawing on it, I fabricated a visual memory which still remains today. It is a reconstruction, however false, and if the event had been real and the memory consequently true, it is very likely that I would have reconstructed it in the same manner because a recall memory does not yet exist (only a recognition memory), in a baby in its pram. [translation of Piaget, 1971, p. 8]

In other words, in Viderman's terminology (1970), construction and reconstruction are inseparable. Freud's archaeological metaphor expresses a similar state of affairs (Haynal, 1983a): points of condensation, like objects in archaeology, speak— see Freud's expression: "*Saxa loquuntur*" (1896c, p. 192). The archaeological object coexists with a whole perspective based on a knowledge of other sources introduced in the "(re)-construction" as an interpretation of the meaning of the object and of its alleged historical context. McCarthy and Warrington (1988) have strongly emphasized the fact that *semantic* information is, like the elements of memory, represented in a *multiple* fashion in the normal brain and that this is linked to modalities of "input" and, thereby, ways through which knowledge is received. (In this context, language is considered as a system of input analogous to others—cf. Marshall, 1984.) For example, if it is assumed that we learn in childhood that whales are fish and later that they are mammals, it may happen, for instance after a brain lesion, that one piece of the information is lost. In normal man, the two would be synthesized, but never perfectly: in certain regressive moments, one or the other may resurface. Such data open new perspectives in psychoanalytic interpretation, one of the functions of which would be to *link* up again our later *verbal*, indeed adult, information with earlier sensory *impressions*. A commonplace example from one of my patients— the traumatic memories of the end of her parents' love-making— had been totally repressed. In one session, the analysand had even gone as far as saying that her parents had probably never had sexual contact (this was, at the same time, an accusation). When she was reminded that her mother had brought her into the world and she could not, as a rational adult, deny that there had been at least one sexual encounter between them, some recasting of her ideas took place, and all of a sudden she recalled a memory of the "primal scene": *two* kinds of knowing were thus brought together again. In more complex situations, putting into the form of adult language an emotional experience whose perceptions remain represented in a mode full of imagery or even more regressive (e.g. tactile) often presents similar impressions that are sometimes very convincing and, indeed, touching.

With regard to the limits of artificial intelligence, observations have frequently been made in this new science that are important and restrictive, reminding us that it ignores bodily sensations, emotions, and engagement in the distinctive character of the human (Dreyfus, 1979). In other words, in its present form artificial intelligence fails to take into account the emotional part of human thinking. For that reason, the psychoanalytic considerations of *representation* and *affect* constitute in their interdependencies (Green, 1973) a further advance in the exploration of human mental functioning. Parallel processing enables two courses to join each other at a given moment— often to the subject's surprise.

Creativity

In Piaget's language, one could speak of a sudden shift in balance towards assimilation and away from accommodation, as is the case, for symbolic play. The individual is generally aware of the problem; there emerges, often quite suddenly, a second series of preconscious ideas that are attainable but not present in the subject's consciousness. This may arise through dreams remembered at the moment of wakening, as in the famous example of Kekulé, who found the solution to a scientific problem through the image of a serpent biting its tail, which led him to the construction of his model of the benzene ring. He describes how he was sitting writing his manual, which was not, however, progressing well, and his mind was elsewhere. He turned towards the fire and fell into a half-sleep. Atoms were flying about in front of his eyes. Small groups remained modestly in the background. His mind's eye, which was accustomed to repeated visions of this kind, picked out larger and more varied configurations. They were in a long series, often more densely assembled. Everything was in a state of flux, writhing and squirming, like snakes. All of a sudden he saw one of the snakes seize its own tail, and the whole configuration swirled around mockingly before his eyes. He woke up in a flash and spent the rest of the night developing the consequences of the hypothesis. His dictum was that men should learn to dream in order to find the truth. To those who do not reflect, it would be offered; they would possess it without any worries. But they should guard against publishing their dreams before examining them in the reasoned light of day (Kekulé, 1890, in Lewin, 1958, p. 46). Poincaré (1913) recounts a similar experience, and Karl Friedrich Gauss (1863–1933, quoted by Rothenberg, 1979, p. 395) declares in his *Memoirs* that the law of induction was discovered on 5 January 1835 at 7.00 in the morning, before he got up. Both Darwin (1892, pp. 42–43) and Pasteur (in Dubos, 1960, p. 114) arrived at their discoveries in the same manner. Charles Darwin attributed his discovery of the selection of species to his reading Thomas Malthus' *An Essay on the Principle of Population* (1798), based on the idea that populations allowed for a greater number of individuals than could reasonably survive and that they evolved through elimination in a fight for existence. Pasteur drew on the model of vaccination in order to the interpret his experiences with cholera. For Kuhn (1962, pp. 174–210), scientific breakthroughs were often drawn from a model in a field other than that of the problem at stake. Notions like Stekel's polyphony of thought

(1924), Rothenberg's Janus-faced thinking (1979), or Koestler's "bissociation" (1964) all seem to me representations of this same model. What is striking from an analytic point of view is that the configurations operating in these discoveries are often fantasms of their creators, linked to personal preoccupations that refer to mysterious links between problems of scientific interest apparently far from the point at issue, and the unconscious affective bedrock of their inner lives. The creation of science is a very human affair—its "detachment", its "objectivity", could prove to be no more than a myth.

CHAPTER SIX

Identification
and countertransference

Still climbing after knowledge infinite
And always moving as the restless spheres.

Marlowe, *Tamburlaine the Great*, 1587

T he understanding of another is so *enmeshed* with an understanding of the self that the most important contributions to psychoanalysis are autobiographical, indeed auto-analytical. The compelling wish to be understood plays a pre-eminent part in what is most advanced, most original, or sometimes most innovative in this field. Self-understanding has an undeniable place in the understanding of others, but it is not without pitfalls. Those problems in the other that do not touch on the analyst's conflictual problems may be felt to have a simple resolution, as being obvious. Conversely, when the problems the analysands bring appear less clear to the analyst, they are likely, in favourable cases, to drive him to understand them better in himself, a motivation that is an important one for the exercise of this profession, unless he is discouraged and frightened by their apparent incomprehensibility.

Psychoanalysts, like everyone else, experience some difficulty in being confronted by unpleasant truths, particularly those that do not correspond to their expectations and ideals. The constituent movement of the mind goes through imitation, identification, introjection, etc., as can be seen in the convergent literature from Freud to Melanie Klein and Piaget. All these semi-synonyms seize upon slightly different mechanisms or precursory stages of that great river that drives the small child to *be like* the Other, mechanisms that are found at one extreme in superficial imitation and at the other extreme in the "holding" of significant aspects of the Other in the processes of mourning. Nevertheless, there are analysts who sometimes neglect the importance of these introjective aspects in the psychoanalytic process, perhaps because they seem to endanger the ideal of the subject's autonomy (non-influence). They accept the introjection of the analyst in his analysing capacity, as the one who participates in the experience *with* the Other, but, probably out of respect for individuality, they have difficulty in examining the profound influences that go from the analyst to the analysand. On several occasions, Freud (1919a [1918], p. 165) emphasized the need to guard against such aspects. The importance of these mechanisms in the "modelling" of the individual in the changes within him, in his internal training during the course of childhood, or in interpersonal encounters such as an analysis seem obvious, however, and, if they should have to be limited, they should be recognized and taken into account.

On the other hand, some aspects have simply been attributed to *"dependence"*—suggesting a rather pejorative connotation—from which the analysand should either be quickly relieved or urged to rid himself as quickly as possible; but in an analytic relationship it is clear that this dependence is created and will last. In ideal cases it diminishes, in fact disappears, towards the end of the analytic relationship, at least on the practical level. Such an intimacy always involves a mutual influence; nor can problems of mutual introjection and identification be avoided. Moreover, this element is frequently found in critical caricatures hostile to psychoanalysis. The resemblance in ageing couples after a long life of sharing might be a well-known similar phenomenon, in some respects, to what happens with the analytic couple.

Identification

Studies made during brief psychoanalytic psychotherapies, which are more favourable to empirical researches than long psychoanalyses, have shown that the changes brought about are directly proportional to several factors. In agreement with Malan (1976) and Marziali (1984), studies by Luborsky, Crits-Christoph, Mintz, and Auerbach (1988) suggest that the more transference interpretations there are, the better the results. According to them, and in accordance with Strupp, Wallach, Wogan, and Jenkins (1963), a liking for the patient and interests and values shared by therapist and patient (and thereby possibilities of mutual identification) are good prognostic factors for change. Furthermore, if the fundamental interests and attitudes of analyst and analysand are broadly matched, mutual identification and understanding are more likely and are more often the origin of desire for change.

On self-analysis

Self-analysis is the understanding of the Other through oneself and from that an understanding of oneself. Whatever the limits—and both Freud and Ferenczi considered it to be unrealizable—it is clear that, paradoxically, self-analysis is a necessity. A whole literature has been written on the topic (Anzieu, 1959/1975/1988).

> My self-analysis is still interrupted. I have now seen why. I can only analyse myself with objectively acquired knowledge (as if I were a stranger): self-analysis is really impossible, otherwise there would be no illness. [Freud, 1950a, letter to Fliess, 14 November 1897]
>
> . . . in self-analysis the danger of incompleteness is particularly great. One is too soon satisfied with a part explanation, behind which resistance may easily be keeping back something that is more important perhaps. [Freud, 1935b, p. 234]

But Freud was *ambivalent* on this subject:

> . . . we require that he shall begin his activity with a self-analysis and continually carry it deeper while he is making his observations on his patients. Anyone who fails to produce results in a self-analysis of this kind may at once give up any idea of being able to treat patients by analysis. [Freud, 1910d, p. 145]

According to this anecdote, the difficulty of self-analysis is "the countertransference". This idea can be found in other thinkers such as Wittgenstein (1931), who said that nothing is as difficult as not deluding oneself ["Nichts is so schwer als sich nicht betrügen]. Some consider that the simple attitude of "evenly suspended attention" (Freud, 1912e) is enough to ensure the best relaxed psychic conditions and to make the associative field of the analyst as free as possible. Others have described "techniques" facilitating self-analysis. For instance, the psychoanalyst can make use of his visual associations with patients (Pickworth Farrow, 1926; Ross & Kapp, 1962). Horney (1942) recommended that the analyst should, in some kind of free or automatic writing, record all thoughts that come to him about several consecutive sessions. Kramer (1959), Myerson (1960), and Engel (1975) have tried to describe examples of analysts' self-analyses. In an effort that can only be called heroic, Calder (1980) has, over the course of 15 years, noted all his dreams, and judged this effective in discovering the central point of traumatization. Ferenczi's Clinical Diary (1985 [1932]) is a unique attempt at exploring the internal associations of an analyst and the emergence of his commentaries, his "theories". For another author, Poland (1984), even a short passage of music can be an interesting signal, enabling the analyst to go further in his associations. Understanding one's own feelings or their activation by the patient and coming back to a more neutral position is an important dialectic within the psychoanalysis. It is not a matter of being perfectionist and purist, indeed pure, in listening, but of not denying the intense work necessary for regaining the analytic stance each time an emotional mobilization occurs. The so-called "neutrality" is acquired by this work; it is one of the great paradoxes of psychoanalysis (Haynal, 1989b). In this way the analyst may induce the patient to bring forward sooner and with less difficulty things he already knows (Freud, 1912e, p. 118).

Marie-Clairette

Marie-Clairette was a young analysand who, from our very first meeting, aroused some anxiety in me. I had the feeling that this analysis could not progress, not succeed. I felt this uneasiness at the preliminary interview, despite the fact that the overall assessment—"cognitive" and emotional, according to the usual internal process—seemed to be rather positive: all the elements required for engaging in an analytic process seemed to be present. So what motivated my unease? At that point I did not know.

The analysand's anxiety showed through only in the form of avoidances: for the most part she remained rather vague. What she said was sometimes enigmatic but generally distant, and, as if to console me, she said from time to time that she thought things were a little better, contrary to the premonitory feelings she had had three or four years earlier, when she had first thought about undertaking an analysis. Her anticipations had therefore been rather pessimistic, and she was surprised when things were better than she had anticipated.

After one of the first sessions, watching her leave my consulting room, I said to myself, "She dances on tiptoe", adding in my inner monologue: "At least she has the lovely legs of a dancer." I caught my remark and at the same time was embarrassed by it; frankly, it seemed a silly thought, a student's or an old bachelor's joke. But I had captured it and perceived it as a kind of distraction, a "self-amusement" to hide my embarrassment and confusion, a response to that hint of seduction, combined with a great effort to keep me at a distance and a "touch-me-not" message, which she directed at me.

I stored my feeling away in my mind, without, for the time being, understanding it. It was only much later that I grasped that, behind the fear of upsetting her equilibrium, there lay the protection against a childhood traumatization that I was ignoring and that I had to go on ignoring for still longer during the analysis; it took the form of a rape, an incestuous secret, which she had sworn to her father she would never mention. And the secret "burst out" after the holidays, when she had the impression that I was making a move towards her. She was frightened because my voice was more relaxed. [The possibility cannot be

excluded that her sensitivity had picked up a grain of truth:
". . . they do not project it into the blue, so to speak, where
there is nothing of the sort already" (Freud, 1922b, p. 226)].
The following night, she had a dream—perhaps to satisfy her
analyst. [How interesting it is to "crack the nut" and find the
meaning of dreams, particularly when it is the reconstruction
of traumatization—what a mutual pleasure!] However, it was
also, at that time, after three years of analysis, a means of
communication that enabled her to say things in veiled terms
through a dream in which everything is said and in the end
nothing is said, because "it is my unconscious that is talking; I
have nothing to do with it. . . ."

Between intimacy and distance

A great deal has been said about psychoanalytic *listening*. It
has been said that free-floating attention is the counterpart of
free associations. What seems important to me is to emphasize
that the analyst's listening is not "linear". According to "multiple inscriptions" in the memory, the analyst never knows by
what path, through what association, in which context a
drawer will suddenly "open". Often there is a feeling that it is
not he, the analyst, his "I", who opens it, but that it opens "by
itself", that the Other releases, from a module of his memory,
associations, unexpected resonances. It is enough that the
analyst is there and that he is open to them.

And yet he cannot contain all those uncertainties; the wish
for a defensive release occurs if the interpretation is there to
stimulate the analysand's associative processes, especially
with regard to excess anxiety or defences. The analyst, too,
needs, through a certain uneasiness, to relocate the relationship—to play a trick, as in a game of chess (Freud, 1913c, p.
123).

Fashionable formulae such as "the psychoanalyst as a person" or "the interactional model of psychoanalysis" grasp only
the superficial aspects of a problem that Freud, Ferenczi, and
others explored in great depth and in cases with multiple
resonances.

The psychoanalytic interaction cannot be conceived as a linear one; that would imply a perfect knowledge of his own psychic mechanisms, as well as the Other's, a perfect "following" of what happens in the interaction. Indeed, it is the *search* for more understanding, the alternation between comprehension and non-comprehension and, thereby, the dialectics of "rapprochement" and remoteness, togetherness and separation, that are the mould for a dual psychoanalytic relationship. This is not a symbiosis, but an alternation between intimacy and distance, which does not turn on itself but reflects meaning to an Other from the past. It is in this sense that the process itself demands the presence of the Other and that the pure interactional model involves the danger of its deteriorating into a "game of Ping-Pong", which is agreeable enough, sometimes amusing, but no more than that. Through the introduction of a third party in the midst of the game the past clarifies the present, the external the internal, dreams and parapraxes clarify the conscious, and beneath all this forms of *integration* can emerge. This integration is not control or mastery, but the realization of potentialities, the promise of a more harmonious life.

The process begins. In the analysis referred to earlier, it began notably at the point when the analyst thought he was beginning to understand *a little* of why he feared failure at the outset. Having realized that the situation of semi-seduction and semi-distancing aroused problems, idiosyncrasies, and hurt in *himself*, he could shoulder the responsibility for the situation and be able to express it to the Other. Quite simple? Yes, explanations *after the event* are very easy, but they presuppose a time of affective receptivity, of *"Erlebnis"* [experience] and understanding. I am talking about understanding in which the aim is understanding of the Other but which happens through oneself. This is how the problem of countertransference and *acceptance* in the widest sense of the term can be seen. Academic distinctions between pathological countertransference linked to the analyst's own problems, the "normal" countertransference aroused by the analysand, and the countertransference determined by the analyst's *modus vivendi* and his relationships in the world outside the analysis are highly sophisticated ones, but they are of no great value in the *situation* of the psychoanalysis. Things do not happen in that way. The first element is an *affect* felt by the analyst; that affect is determined by what he has just heard and by what he is because of his past, through his vulnerability and the circumstances of his present life. What matters in this associative chain is that he is finally able to reconstruct the *affect* and the *fantasmatic contents* that are linked up with the analysand, to whom they also belong. It is a *strategy* decision and not an ambition to expose "the whole truth". Besides, the analyst will take, and take in part of this truth for himself. . . .

The *process* is the work of two protagonists; it unfolds in time through the contribution and the development of both the analyst and the Other.

In the evolution of psychoanalysis, the exaggerated phenomena were the first to be grasped. The description of exhibitionism and sadism in their radical, extreme form paved the way for subsequent investigations of similar attenuated phenomena. This was so in the case of countertransference. The most basic, most obvious affects that were mobilized, those that were the most disturbing, were the ones that attracted attention to the importance of the inner strivings of the analyst. Other historic factors, aimed at questioning the authority of professionals, resulted in the analyst no longer passing for an exceptional being but for a human being like others, with his own feelings and emotions. The general development of the profession and the refinement of knowledge have made it possible to investigate the finer points of the field thus opened up. From this point of view, interpretation is a *projective* displacement, since the analyst succeeds in understanding the Other in an introjective move, and it is in a projective move, i.e. centred on the Other, that he will express it. Canetti described the importance of learning to *listen*. Throughout, at every moment, *everything that is said* by anyone offers a dimension of the world that up to then *had not been suspected*, as it is an alliance of language and men, in all its variations, which is perhaps the most meaningful, and certainly the richest. This form of listening is *not possible without the renunciation of personal reactions*. As soon as one has set in motion what could be heard one stops, because one must listen without allowing interference from any kind of judgement, indignation, or enthusiasm.

He goes on to say that for a long time there was little idea of the *fullness of the material* that had been accumulated. What was felt was only a thirst for a kind of language that was as explicit and as clearly delimited as possible, a wish to be able to take in one's hand as an object what suddenly occurred in the head provided that their relations, whatever they were, could be recognized so well that one could not help *repeating them in a loud voice* (Canetti, 1967, II, p. 234).

Harold Stewart (1987) distinguishes several types of transference interpretation:

1. interpretations aimed at understanding the drive/anxiety/ defence conflict between the patient and the analyst, in an anaclitic type of object relationship;
2. an extension of the above in the sense of interpenetrating simultaneous conflicts of a similar or complementary nature;
3. interpretations aimed at understanding the sensitivity and vulnerability of both patient and analyst in a narcissistic-type relationship;
4. through extension of the above, the interpretation of the failure to reach such an understanding;
5. interpretations aimed at understanding the ambience or seeing the humour or the present mood in the relationship between analyst and analysand;
6. interpretations aimed at understanding the analysand's unconscious response to the analyst's interpretations.

A committed detective?

The psychoanalyst engages in enormous inner activity: feeling (receiving the affective message), understanding, while deferring his direct response but, paradoxically, retaining some spontaneity (in Canetti's sense when he refers to not being able to refrain from repeating [his understanding] in a loud voice). Construction, creation of models, their abandon, the enrichment by new ideas, the construction of new models, are all means of deepening the understanding, starting from hints and going back to them, like a detective, a Sherlock Holmes:

> The task of the therapist, however, is the same as that of the examining magistrate. We have to uncover the hidden psychic material; and in order to do this we have invented a number of detective devices. . . . [Freud, 1906c, p. 108]

> And if you were a detective engaged in tracing a murder, would you expect to find that the murderer had left his photograph behind at the place of the crime, with his address attached? Or would you not necessarily have to be satisfied with comparatively slight and obscure traces of the person you were in search of? So do not let us underestimate small indications, by their help we may succeed in getting on the track of something bigger. [Freud, 1916–17, p. 27]

In order to be able to disengage oneself, to hold out, to function, the analyst will interpret, at the very least in order to pin down his own role in the ongoing relationship.

How far does his commitment go? As far as his dreams? If the problem of his work penetrates his dreams, it may serve as an indication of an important disquietude with regard to his analysands (Zwiebel, 1985, pp. 87 et seq.). The analyst's extensive participation is made up of his personal vulnerabilities, whence the importance of understanding his dream—or his slips of the tongue and parapraxes—in relation to the situation that triggered them or fed into them. We can only really understand our own distress; the Other's pain is comprehensible in so far as we can identify with it. The force of the images behind it arouse in us the reminiscence of similar constellations, and it is *this analogy* that enables us to understand. Without such a bridge we are lost. Explanations that manage to articulate the nature of the problem, especially in another relationship, help to "sort it out". That is the value of assisted listening through supervision or monitoring; another ear, one more to listen, may sometimes hear what the analyst does not manage to hear by himself.

This exploration of the inner life often consists of putting forward an idea, a concept, like that of countertransference, and *then* looking at its implications and limits.

It is a process that is more typical of hermeneutics than of deductive epistemology. But this does not mean at all that the symbols and their exploration do not give any information about the semantics involved—the *field of meanings* and what is *behind* it in the evolution of man, in his history and in his growing. It is at this second level that the analyst can grasp laws, lines of force of that evolution. It is a method adapted to the field from the beginnings of a science of man about himself, developed throughout the twentieth century. It starts with the affective communication *captured* in the countertransference and the reflections that it provokes.

The countertransference thus becomes an exceptional tool for analysing the Other (Little, 1951). Its function is not to induce the analysand to indulge in confessions that would only serve to relieve the analyst of some guilt or unease, which is not the aim of analysis. Empathy presupposes a certain sympathy, a receptivity [*"Stimmung"* or *"Gestimmtheit"*]. But sympathy in itself is not empathy. Empathy is something more; it is *"Einfühlung"*, the capacity to put oneself in the Other's skin, a capacity that is furthered by sympathy.

When *difficulties* are experienced in identifying with the basic motivations or deepest desires of the Other, an embarrassing countertransference situation is produced: grandiose, aggressive, or destructive ideas—especially when they appear in the guise of motivations for analytic treatment (for example, "grandiosity")—make any complicity, any therapeutic alliance based on them very difficult. Accepting them, even provisionally, in order subsequently to set about recasting them can, in fact, from an affective point of view create considerable obstacles, especially if the analyst fails to be sufficiently aware of the "game" being played.

If, with Freud (1925d [1924], pp. 10–11), we consider transference as full of unrealizable expectations ("He takes me for his father"), like, for example, the transference of twins, in which the other is taken for an ideal or identical being or, on other occasions, as "a poor chap", we may also recognize in it the *reproduction* of a psychosocial situation, a possible line being envisaged with biological determinisms (Gardner, 1987). Psychoanalysis offers a broad general framework in which the problems of modern man can be rethought in the context of information furnished by the different sciences. At the present time, many such broad frameworks are empty, awaiting data and information that these sciences may well supply.

Countertransference/identity of the psychoanalyst

Through the deepening of this conception of an encounter between two equal parties on a sustained course with a profound and important engagement by both, psychoanalysis has taken the path of a personal task on the part of each of the protagonists. It is an artistic task, too, in some ways—a "creative" one, even—for the two partners the aim being a renewal, an opening up of the whole personality but at the same time an integration, at the level of consciousness, of the different parts of the personality in all its richness. There is no doubt, with such ambitions at stake, that the original aims of psychoanalysis, as Freud described them in his cases of Katharina (1895d) or Little Hans (a psychoanalytic "influence" on a boy through his father—1909b), have been totally surpassed. Furthermore, there is a clear bifurcation to be seen here. On the one hand, psychoanalytic treatment as therapy is moving towards brief psychoanalytic psychotherapy (Balint, Malan, Sifneos, etc.), towards therapies practised by doctors or other professionals in the community (social workers, marriage counsellors), and towards the most varied "derivatives" of psychoanalysis (Alexander and others). On the other hand, a more and more profound psychoanalysis is developing in the spirit of a potentially infinite encounter, making radical demands on both parties. This second perspective raises the question of knowing whether what we are talking about is still "*treatment*", in Freud's sense, or whether it is an enterprise that goes a great deal further, perhaps more in the spirit of an existential encounter.

It is a subject that excites strong feelings because the identity and *professional ideals* of each practitioner are being challenged. For some, psychoanalysis is what has become of it in the wake of a long development; for others, psychotherapeutic treatments inspired by it are a legitimate continuation of the treatments practiced by Freud—for example, on Gustav Mahler (Jones, 1955)—and his many followers, Winnicott, and others.

Clearly, therefore, all these questions seem to be linked to "countertransference" in the widest sense of the term—namely, to the psychoanalyst's professional *identity*, to his *ideals*: what he wants to offer, how he wants to work, and what benefit he can derive from it. There are analysts who "don't like" brief psychotherapy: by plunging into analyses, quite obviously some of their undoubtedly legitimate desires are satisfied—in my opinion, not only their curiosity or the wish to help, but also their need to continue their own analyses in this deep identificatory relationship, which is built up through long treatments. Those who are inclined to undertake quicker treatments, indeed brief psychotherapies, often have a greater need to *help in the shorter term*, with perhaps more circumscribed results. A not inconsiderable number of these latter also turn their attention to the scientific question of factors of change on the one hand and "psychodynamic constellations" on the other. No doubt the majority of psychoanalysts organized in the international professional movement have taken rather the first stance; and no doubt, too, in the ecological conditions and demands of the public, there are differences between continents (between North America and Europe, for instance, or, within the latter, between the northern and Latin countries). Brief psychotherapy and its concomitant concerns are more readily acknowledged in North America and in northern Europe. It is clear, however, that it is a historical development that has resulted in such a differentiation influenced by understandable factors. In the end both rely on the same tradition. In the Father's house are many mansions, as it has evolved—constructed and reconstructed—historically.

Repercussions of communication
on the subject: change

Freud did not lose interest in problems of *change*—a succession of theories on the theme bears witness to that. But for him it was a secondary problem, a question annexed to his main preoccupation of understanding mental functioning in its *entirety*. Thus each theoretical revision had some repercussion consequent on each stage of his thinking.

> By explaining things to him, by giving him information about the marvellous world of psychical processes into which we ourselves only gained insight by such analyses, we make him, himself, into *a collaborator, induce him to regard himself with the objective interest* of an investigator, and, thus push back his resistance, resting as it does on an affective basis.
> One works to the best of one's power, as an elucidator (where ignorance has given rise to fear), *as a teacher*, as the *representative for a freer or superior view* of the world, as a *father confessor*, who gives absolution, as it were, by a continuance of his sympathy and respect after the confession has been made. [Freud, 1895d, p. 282, italics added]

These two quotations show Freud trying to find a collaborator in the patient, in the framework of a scientific research jointly undertaken.

Five years later he takes into consideration the *resistance* to change; the concept of recall is replaced by that of representation and then by that of fantasy:

> The main result of this year's work appears to me to be the surmounting of fantasies; they have indeed lured me far away from what is real. Yet all this work has been very good for my own emotional life; I am apparently much more normal than I was four or five years ago. [Freud, 1950a, letter to Fliess, 2 March 1899]

This "long way from reality" led him to modify his collaboration with the patient. *Fantasy*, and then, in the case of Dora, *transference* were to be the important levers, enabling the desired change to be achieved:

> Transference which seems ordained to be the greatest obstacle to psychoanalysis, becomes *its most powerful ally*, if its presence can be detected each time and explained to the patient. [Freud, 1905e, p. 117, italics added]

This raises the problem of repetition and working-through in the transference, which was considered at that time as "an

artificial illness which is at every point accessible to our interventions" (Freud, 1914g, p. 154), leading to the libidinal
cathexis of "a new object, and the libido is liberated from it"
(Freud, 1916–17, p. 455). It is a sort of *"renewal"*—probably the
starting point in Freud of Balint's reflections on the concept of
"new beginning" (Balint, 1933). Transference, as always, plays
a very important role:

> The main instrument, however, for curbing the patient's
> compulsion to repeat and for turning it into a motive for
> remembering, lies in the handling of the transference. We
> render the compulsion harmless, and indeed useful, by
> giving it the right to assert itself in a definite field. We admit
> it into the transference as a *playground* in which it is al-
> lowed to expand in almost complete *freedom* and in which it
> is expected to *display* to us everything in the way of patho-
> genic instincts that is *hidden* in the patient's mind. [Freud,
> 1914g, p. 154, italics added]

(The expressions "playground", "complete freedom", "display",
and "hidden" are strikingly reminiscent of Winnicott.)

One senses Freud's groping with the concept of transference,
foreshadowing, perhaps, notions of "Gestalt" and "schema"
in their later meaning. He also anticipates the metaphor of
integration by saying that after eliminating the resistances, the
mind

> grows together. The psychosynthesis is thus achieved dur-
> ing analytic treatment without our intervention, automati-
> cally and inevitably. We have created the conditions for it
> by breaking up the symptoms into their elements and by
> removing the resistances. It is not right to think that some-
> thing in the patient has been divided into its components
> and is now quietly waiting for us to put it somehow together
> again. [Freud, 1919a [1918], p. 161]

But after the great discoveries and the courageous hypotheses, the weight of failures and difficulties is experienced without any concrete and convincing answer. Freud refers to
generalities in a rather speculative vein: the difficulties are also
attributed to the adhesiveness [*"Klebrigkeit"*], the viscosity—or
even to the exaggerated mobility—of the libido, to an excessive
fixation (1916–17, pp. 346), to the role of the death instinct

(1920g), the wish for a penis, and the masculine protest (1937c, p. 252).

Obviously, it seems necessary to rethink and bring to light the theory of change. Awareness of affective communication and the integration of its meaning, to my mind, opens a new perspective on the elements already advanced by Freud. Transference—the affective aspects of the past, caught in the countertransference and translated into a verbal language rendering them coherent—enables the construction of a model of integration that, in its turn, verifies its validity.

PART TWO

THE HISTORICAL DIMENSION IN THE CONSTRUCTION OF PSYCHOANALYTIC THEORY

CHAPTER SEVEN

Freud in the cultural tradition

> The best way of understanding psychoanalysis is still by tracing its origins and development.
>
> Freud, 1923a, p. 235

Rationalism and romanticism

I n the eighteenth century, rationalism reigned triumphant: it was the age of Enlightenment, of Voltaire, of the Encyclopaedists, and of the German *"Aufklärung"* and its political expression in the French Revolution. Moreover, it was the time when the natural sciences came into being, the age of Lavoisier, one of the creators of modern chemistry and the man who introduced the metric system, a symbolic figure and one whose own destiny was linked to the Revolution.

The Germanic countries

The next epoch saw the reaction, especially in the Germanic countries, which Reason was bound to provoke. At the turn of the century, these regions were faced with what can properly be

117

called *romanticism*. As a conception of being, romanticism is of interest to our own profession—romanticism as it is expressed by the ageing Goethe in his "Colour Theory", by Novalis, Hölderlin, and several other German writers. "Man must be considered in his entirety." Illness does not happen as a chance "accident" but is the conveyor of a precise meaning, which may be positive; it is an escape from control, an initiation. Such ideas are strongly emphasized by the representatives of this movement. The performance par excellence of the romantic death "giving a meaning to the whole of life" is Byron's in 1824—an apotheosis crowning a broken life and at the same time a gesture expressing the aspirations of a whole generation towards a better life and liberty.

In the romantic movement, Friedrich Schelling's [1775–1854] "Nature philosophy" (1797) puts emphasis on notions such as "global" and "meaning" and puts forward new ones—"organism", for instance. This was, however, considered by many scientists as retrograde and unscientific: the term "organism" is specifically intended to designate that "entirety", which is more than the sum of its parts and is therefore different from those parts. Goethe, in his "Colour theory" or especially in the "Metamorphoses", excited by the understanding of nature, considers that counting [*"zählen"*] is quite unworthy of science. What is important for him is the wish to grasp facts intuitively, refusing to examine all their minutiae and to subject them to the rigours of mathematics—in other words, the refusal of everything that characterizes the natural sciences of the nineteenth century.

The successors of the nature philosophers are the ancestors of alternative medicine. Hufeland, greatly encouraged by Goethe, is the founder of *macrobiotics*. Around 1790, the Saxon Samuel Hahnemann was struck by the fact that quinine given to a healthy individual produces effects identical to those that quinine cures in a sick person (Gusdorf, 1984). Generalizing from this observation, he did similar studies on the effects of mercury, digitalis, belladonna, and other pharmacopic drugs, reducing the doses considerably. The change in quantitative scale, far from diminishing the effects, seemed, on the contrary, to increase their potency. By producing symptoms similar to those of the illness, the illness was suppressed, not provoked (Gusdorf, 1984). This is the founding principle of *homoeopathy*.

We are thus witnessing at the end of the eighteenth century and the beginning of the nineteenth the moment of the development of the nature sciences, the dawn of alternative medicine. From the point of view of developing scientific medicine, it is the beginning of "charlatanism". Bircher's diet (from which *"birchermuesli"* is derived) is part of the same movement, along with the Anthroposophical Society, an esoteric movement created in Vienna at the end of the nineteenth century by Rudolf Steiner, in the belief that the *doctor's spiritual* relationship with handicapped or mongol children is an essential factor in *his* own development. (Steiner believed in reincarnation, affirming that in a previous existence Karl Marx had been a French warrior who pillaged his neighbours. According to Steiner, he discovered one day on his return that a nobleman had helped himself to his castle and land and had become his vassal. Master and slave were reincarnated as Marx and Engels—Wilson, 1988.)

Friedrich Mesmer [1734–1815] was interested in hypnosis, in dreams and somnambulism; in all the phenomena that later came to be referred to as hysterical. *Magnetism* offered a new understanding of the "physical" ideas of the age. The first steps towards psychotherapy were then taken at the boundary of the "romantic"—following the example of the Swabian doctor, Justinus Kerner [1786–1812]—and the "scientific". In this same tradition, Freud was also to practise hypnosis, to treat hysterias, and, most importantly, to seek a scientific explanation for romantic themes. The dream, hysteria, sleep-walking, the *Doppelgänger*, and "the uncanny" of E. T. A. Hoffmann would be at the centre of his preoccupations. As an inheritor of the spirit of the Age of Enlightenment, Freud could be said to have introduced the great romantic themes into the *"field* of the sciences".

Schelling's natural philosophy of nature was closely followed by the founding movement of *scientific medicine*. His pupil, Johannes Müller [1801–1858], was to take the first chair of physiological anatomy in Berlin. As a teacher at the university there from 1833, he formulated the theory of "the specific energy of nerves", which was to revolutionize neurology and give birth to the great line of German physiologists: Du Bois-

Reymond and Virchow were his pupils, and Helmholtz developed out of the same circle. The first generation of Müller's emulators founded the "Berliner Physikalische Gesellschaft" [Physical Society] in 1845 with the help of Du Bois-Reymond, Helmholtz, Ludwig, and Brücke (Freud's venerated teacher). In 1845, Robert Mayer, a trained doctor, put forward the *principle of energy conservation*, which considered the vital processes from the angle of the transformation of force or matter. Julius Liebig, the founder of organic chemistry, was one of the first at the beginning of the nineteenth century to hold a chair in chemistry. Wilhelm Wundt, a pupil of Helmholtz, was the father of *scientific psychology* (psychophysiology), following Lotze's work. He was responsible for the "physicalization" and experimental evaluation of psychology.

It is said that Emile Du Bois-Reymond, Helmholtz, and Brücke swore, with the symbolic gesture of an oath-taking, that they would acknowledge no explanation in the study of nature other than that provided by physio-chemical forces, such was the enthusiasm of their commitment. The historical—and more prosaic—truth lies in a letter from Du Bois-Reymond to his friend Helmholtz written in 1840. There is no trace of an oath.

The majority of these scientists came from a generation of *doctors* who, in order to get closer to the enigmas of physiology, devoted themselves to the physical science and became the founders of it. This approach was a determining factor in the orientation, methodology and discoveries of the medicine of the time. It is in this sphere that ideas arose which today are still a reality.

1. Johannes Müller's *vitalism*: the last representative of "Nature philosophy", he maintained that the phenomenon of life, over and above its physico-chemical components, assumes the existence of a vital force (whence the term "vitalism"), which is of a different order.

2. The notion of *reductionism*, which covers the other aspect of the discussion: viz. whether or not the biological phenomena are reducible to physico-chemical forces. Towards 1860 this problem was based in the context of Wundt's work on psychology: could the phenomena of psychology be grasped by reducing them to biological forces?

Inaugurated first in Berlin, the chairs of medicine began to multiply in the universities to such an extent that henceforth one could talk about "official" medicine. This rested primarily on the new physiology, then, through Virchow's work, on the new pathology, and, finally, on the expansion of discoveries on histology and bacteriology.

France

In *France* at the same time, at the beginning of the nineteenth century, Auguste Comte's *positivism* appeared on the scene, along with experimental medicine represented by Claude Bernard. (Among his English admirers, John Stuart Mill and Jeremy Bentham figured. It is interesting that Mill's works were translated into German by Sigmund Freud. . . .) With his *Introduction to the Study of Experimental Medicine*, which appeared in 1865, Bernard brought biology into the field of scientific determinism. Physics and chemistry became the bases of physiology, the concept of "vital force" having no place in a scientific explanation of the phenomena of life. His idea of an "internal milieu" brings new clarification to physiological processes. Calling on his knowledge of the organic chemistry of his time, he specified the role of the pancreas in digestion, the glycogenic function of the liver, and vasomotor regulation, and he indicated the complexity of metabolism. This is also the age of the fundamental discoveries of the morphological elements carrying nervous excitation (bear in mind the heroic debates on the contiguity or continuity of nerve cells).

These antecedents played an important part in what was to become *psychoanalysis*. Freud was the direct *pupil* of Brücke and, indirectly, of Helmholtz. He was influenced by Herbart's psychology, in which there were already the concepts of representation, repression, resistance, energy, inhibition, and displacement.

It should be recalled that the unconscious as Freud described it has, itself, a long trail in the history of ideas (Whyte, 1960). The physician and nature philosopher Gustav Theodor Fechner founded *experimental psychology*: his work *Elemente der Psychophysik* [Elements of Psychophysics], published in 1860, was its point of departure. Leipzig, his adopted home town, soon became the centre of this new science. His pupil, Wundt, opened the first Institute of Experimental Psychology there in 1879. Fechner was the author of such concepts as the "pleasure principle" and the "tendency to stability" (the forerunners of Freud's constancy and repetition principles), as well as those of "mental energy" and the "topography" of the mind.

Freud was impressed by these ideas. A number of concepts, such as those of "nervous excitation" and "energy displacement", had a profound influence on his work, and his only ambition was to create a science continuing the ideas of his masters and the natural sciences of his time.

Moreover, his psychiatry teacher, Meynert, was motivated by the wish to explain the human mind in terms of anatomy and physiology, the main thread of his hypothesis being the existence of a functional antagonism between the cortex as the seat of representations on the one hand, and the cerebral trunk, seat of the affects, of involuntary, instinctive, and automatic actions on the other. This antagonism is found in Freud in the "id" and "ego", where, under the influence of romanticism, his interest in the instincts found its place. His intention was to create a discipline of the *order* of *"Naturwissenschaft"* (the science of nature—Freud, 1940a [1938]). Its *object of study*, however, was none other than some of the great romantic themes: dreams, the uncanny, the double, parapsychology. The aim of his great initial project, proposed in his "A Project for a Scientific Psychology" (1950a [1895]), was to *clarify through science* what the *Romantics* and their predecessors, the writers and artists (his beloved Shakespeare and other poets and novelists), *sensed* intuitively. Moreover, this "Project" was addressed to neurologists, which says a great deal.

Freud attributed his vocation to the reading of a work on "Nature" which at the time was attributed to Goethe [". . . and it was hearing Goethe's beautiful essay on Nature read aloud . . . by Prof. Brühl . . . that decided me to become a medical student" (Freud, 1925d [1924], p. 8)], but which is now known to be the work of a romantic theologian called Tobler.

His readiness to dissociate himself from every off-shoot of "Nature philosophy" and what academic medicine considered as charlatanism, explains his reserve towards Groddeck and his tendency to dissuade his students from plunging into psychosomatic medicine "for didactic reasons", as he wrote to von Weizsäcker in 1932 (Uexküll, 1986). He felt that their training should be strictly on the basis of a psychological methodology, and he feared that somatic illnesses would lead them into weak methodological coherence. It should not be forgotten that the very name of his method ("analysis") recalls analytic chemistry and inductive and analytic methods in the chemistry and physics of his day. [Note that Freud used the word "psychoanalysis" for the first time in an article written *in French* entitled "The Heredity and Aetiology of the Neuroses" (Freud, 1896a).] Until the arrival of thermodynamics, particularly entropy, and later, in 1905, Einstein's theses, one single grand synthetic idea had appeared in the natural sciences: Darwinism introduced a new vision of man in the name of science, a revolutionary paradigm. Freud was powerfully influenced by its stamp.

There were other influences also. At the turn of the century Vienna was an extraordinarily stimulating milieu. Wittgenstein was questioning there what is really said when we speak—that is, the true meaning of our words. The Freudian question, or at least one of them, was similar: what do we say when we speak; what lies behind the words?

Scientific intention

Freud writes, in his own characteristic manner: ". . . it still strikes me as strange that the case histories I write should read like short stories and that, as one might say, they lack the serious stamp of science" (Freud, 1895d, p. 161)—a sentence that could mean: "I am neither a novelist nor a romantic. I am, and I want to be, a man of science. If my observations of patients read like novels, it is probably because of their subject matter. My aim is to make a science of it." Elsewhere he declares that the poets are his masters (Freud, 1920g, p. 44; 1925d [1920], p. 32)

If we add his affirmation that psychoanalysis is a "natural science like any other" (Freud, 1940a [1938], p. 158), his meaning seems to be: "Even if Shakespeare and Goethe were my masters, I have not written plays, like Schnitzler [his curious 'double' whom he resisted really knowing], but scientific works." For the same reason he customarily emphasized that he had nothing in common with the philosophers. Even the concept of the "id" was—in his own words—not taken directly from Nietzsche, but through the mediation of Groddeck ["I propose to take it into account by calling the entity which starts out from the system Pcpt. and begins by being Pcs. the 'ego', and by following Groddeck in calling the other part of the mind into which this entity extends and which behaves as though it were Ucs., the 'Id'" (Freud, 1923b, p. 23)]. His attitude seems to be: "I do not write dramas, like Schnitzler, nor do I wish to do philosophy; I want to establish a scientific construction."

This scientific construction, his ambition, was and remains the secret as much of his successes as of the hostility he attracted. His success comes from the fact that he effectively produced a *model* enabling modern man to *reflect* and to *talk* about himself and his fellow men. The spread of psychoanalysis and expressions such as "repression" and "Freudian slips" into our teaching (for instance, ideas of freedom, especially from too repressive authoritarianism, etc.) either come from Freud's ideas or can at least be traced back to them. (What is attributed to Freud is sometimes very different from what he actually expressed in a much more complex fashion.) For the rest, the conceptual tool that he created has had a far greater cultural impact than the treatment technique that he put forward and with regard to which he manifested some scepticism, particularly towards the end of his life (Freud, 1937c, p. 230).

A large part of what Freud and his school developed is common knowledge today. Every parent observes, with his or her child, facts that hitherto would have gone unnoticed: his affective needs, his oral play, the importance of identification in learning, the period of giving and taking, of discovering the law, etc. This is even more apparent in psychiatry. The language of phobias and obsessional neuroses is Freud's language. Today's predominant system of classification of illnesses—the American DSM-III—which purports to be scientifically descriptive and objective, nonetheless implies, behind the obsessional neuroses and phobias, a pathology of anxiety, which inevitably harks back to Freud. Today, our thinking on dreams, mourning, and love is interwoven with hypotheses and viewpoints that issue from his ideas.

Is his general system an empirical science, or a provisional outline, the success of which was above all *ideological*—in the sense that it offers some orientation in this world of ideas and the possibility for reflection on the internal world of man in a manner corresponding to a scientific, lay image rather than a religious one.

Freud certainly endeavoured to take account of the empirical facts and to gather them; but above all his aim was to transmit a more far-reaching image of his understanding of man. When, as often happens, he is compared with some other men of science, like Darwin, Einstein, and sometimes Marx, it is impossible not to see that in reality these men do not owe their prestige to their personal researches but to the creation of a new image of man or of the world, founded on their respective sciences—Darwin with the theory of evolution, Marx with his thesis of man's determination by economic facts, Einstein through a new vision of physical relations, and Freud through a new vision of man. Freud was aware of having contributed, in the name of science, to the creation of new myths in the place of the old ones: "The theory of instincts is, so to say, our mythology" (Freud, 1933a, p. 95). But its extraordinary synthesis has *impressed* the world—or provoked some hesitation in accusing it of being precocious or unproven. In any case, the scope of the concepts and what he calls his mythology have proved to be fruitful from the viewpoint of *heuristics*. The fact that Darwin and Einstein are mistaken in some details does not detract from their historical performances, nor from their influence on later researches and on the conception of the world in general.

Wittgenstein's (1931) declaration that the true merit of a Copernicus or a Darwin is not the discovery of a true theory, but of a new and fruitful way of looking at things, applies equally well to Freud. Referring to himself, Wittgenstein explains:

> I believe that my originality (if that is the right word) is an originality belonging to the soil rather than to the seed. (Perhaps I have no seed of my own). Sow a seed in my soil and it will grow differently than it would in any other soil. Freud's originality too was like this, I think. [p. 36e]

He even thinks that it will take many years until we lose our subservience to psychoanalysis.

"Our point of departure is practice—the psycho-analytic treatment", and Freud characterizes it as "a conversation between two people equally awake" (1904a, p. 250). "Nothing takes place between them except that they talk to each other" (Freud, 1926e, p. 187).

A year later, he added:

> Indeed, the words "secular pastoral worker" might well serve as a general formula for describing the function which the analyst, whether he is a doctor or a layman, has to perform in his relation to the public. We who are analysts set before us as our aim the most complete and profoundest possible analysis of whoever may be our patient. We do not seek to bring him relief by receiving him into the catholic, protestant or socialist community. We seek rather to enrich him from his own internal sources, by putting at the disposal of his ego those energies which, owing to repression, are inaccessibly confined in his unconscious. [1927a, pp. 255–256]

> We are aware, however, that such a definition is inadequate. To the exchange of words must be added the frame of reference of the one who is listening—both recognition of unconscious phenomena, of the transference and of resistance which are its cornerstones. [Freud, 1914d, p. 16; 1923a [1922], p. 246]

So Freud can write to Groddeck: "Whoever recognizes that transference and resistance are the pivot of treatment belongs irrevocably to our 'savage horde'" (1960a, letter to Groddeck, 5 June 1917). He mentions a number of rules designed to facilitate the encounter he calls "psychoanalysis". He considers the possibilities for development in technique as an open question, and he begins to see that technical problems are linked to the person of the therapist, to his own difficulties, for example in being in touch with his own feelings (Freud, 1910d, p. 145). He clarifies this: "I think I am well advised, however, to call these rules "recommendations" and not to claim any unconditional acceptance for them" (1913c, p. 123). He even criticizes those of his students who might have the tendency to take his suggestions literally or to exaggerate them (Freud, 1963a, letter of

22 October 1927). He also writes: ". . . this technique is the only one suited to my individuality; I do not venture to deny that a physician quite differently constituted might find himself driven to adopt a different attitude . . . " (1912e, p. 111). The unanswered—and perhaps unanswerable—questions about technique thus fired controversies over psychoanalytic theory and *practice* and the links between them (Haynal, 1987a, p. 2).

> The theory of repression is the cornerstone on which the whole structure of psycho-analysis rests. It is the most essential part of it; and yet it is nothing but a theoretical formulation of an experience which may be reproduced as often as one pleases when one undertakes an analysis of a neurotic without resorting to hypnosis.

With this statement Freud is trying to construct a science of phenomena he is to call "neurotic". He concludes that his researches have led "inevitably to the view of unconscious mental activity which is peculiar to psychoanalysis", but he hastens to add

> which distinguishes it quite clearly from philosophical speculations about the unconscious. It may thus be said that the theory of psycho-analysis is an attempt to account for two striking and unexpected facts of observation which emerge whenever an attempt is made to trace the symptoms of a neurotic back to their sources in his past life: the facts of transference and resistance. Any line of investigation which recognizes these two facts and takes them as the starting-point of its work has a right to call itself psychoanalysis, even though it arrives at results other than my own. [1914d, p. 16]

A statement in the spirit of the sciences, without the shadow of a doubt.

Closed system—unfolding science

In what Freud conceived, part is his great success; another part will probably not stand the test of time. The question occupying us here, however, is whether or not he actually created a science. First of all, if he did not create a "natural science like any other", his ideas nevertheless established a *scientific tradition*, which he defines thus (Freud, 1923a [1922]):

> The assumption that there are unconscious mental processes, the recognition of the theory of resistance and repression, the appreciation of the importance of sexuality and of the Oedipus Complex—these constitute the principal subject-matter of psycho-analysis and the foundations of its theory. [p. 247]

In other words, he designates repression outside consciousness of painful content, the evaluation of the importance of sexuality in human conflicts, and resistance when these same contents may become conscious as the fundamental hypotheses of psychoanalytic science. They offer a frame of reference permitting investigation of the human mind, which ought to be able to follow scientific principles (Freud 1933a):

> Psycho-analysis, in my opinion, is incapable of creating a *Weltanschauung* of its own. It does not need one; it is a part of science and can adhere to the scientific *Weltanschauung*. This, however, scarcely deserves such a grandiloquent name, for it is not all-comprehensive, it is too incomplete and makes no claim to the construction of an accomplished system. [p. 182]

Freud always thought that in the scientific field, truths are only provisional. Thus, in *Beyond the Pleasure Principle* (1920g), he wrote:

> Biology is truly a land of unlimited possibilities. We may expect it to give us the most surprising information and we cannot guess what answers it will return in a few dozen years to the questions we have put to it. They may be a kind which will blow away the whole of our artificial structure of hypotheses. [p. 60]

If, therefore, he did not create a science along the lines of Helmholtz, Brücke, and others—a natural science in which the concepts of energy and drive retain the same meaning as in the original biological context—those concepts will be increasingly "metaphorized" and their original meaning widened. Is it, as it is often rather dramatically said to be, an "epistemological break"? I do not think so. It is as much a break as a continuity: Laplanche (1970, 1982) considers that a "false" anatomo-physiology is at the basis of the Freudian derivation of these concepts. If so, then there is no doubt that, in pursuing such a path, the Freudian genius gave new signification in a new context. Freud was nostalgic about the original context, but what he accomplished conceptually he did despite that nostalgia, with the sure intuition of a scientific creator. As for the success and value of his system in its entirety, it is not in the details that it should be judged. I shall therefore return to it later, offering a wider, more global approach.

Freud tried to create a theoretical scaffold for a subject about which he said: ". . . for the moment we have nothing better at our disposal than the technique of psycho-analysis, and for that reason, in spite of its limitations, it should not be despised" (Freud, 1940a [1938], p. 182).

Without a doubt, his ambition and his attitude were both *scientific*, even if his methodology was based on "guesswork" and "speculation". (According to Guttman, Jones, & Parrish, 1980, he uses the word "guess" 218 times and its derivations "speculation", etc., 108 times.)

Within this scientific tradition, his statements have been continually revised by successive generations of psychoanalysts. For this reason, it is not fair to compare the first pronouncements with present-day conclusions. Dieter E. Zimmer (1986), for example, compares Freud's original statements with today's science and concludes that they are outmoded, which is like comparing Darwin or Lamarck or Maxwell with contemporary physics and saying that much of what those great minds created has been superseded. Psychoanalysis, like all other sciences, has developed, and it has done so in an interdisciplinary collaboration with neighbouring sciences. Many Freudian theses are no longer retained by psychoanalysts today or, if they are, only in a revised form. Revision—the discovery of new facts and the construction of new hypotheses—is a daily feature in the sciences; why should it be otherwise in psychoanalysis? Freud can be held responsible for his basic hypotheses, such as the theories of repression and intrapsychic conflict, for instance. It is obvious that his scientific models should in part be revised. If he based his conceptions on those of the natural sciences, it is considered today that he explained psychic processes through metaphors the value of which is descriptive—without concepts having necessarily been found that are coextensive with adjoining sciences (energy with physics and physiology, drive with biology). Moreover, these concepts are not all situated at the same level in methodology and epistemology.

But let us pursue the history of the psychoanalytic ideas of which many empirical investigations were the outcome. To mention only a few of the most important, Piaget's clinical method resulted from his encounter with psychoanalysis. Wondering why in the Binet-Simon test some children managed to give false responses, and what function this had, he raised a typically psychoanalytic question. Works like that of Paul Kline (1972) and Seymour Fisher and Roger P. Greenberg (1977) showed that some parts of psychoanalytic theory contributed to a better understanding of child development. Spitz and Bowlby demonstrated the importance of attachment and that one of the chief causes of human suffering resides in sudden early separations of the child from the caring person. More recently, Daniel N. Stern has brought to the fore basic interactional models which play a fundamental role in the constitution of the human mind and especially of the sense of self in the baby and the young child.

The controversial libido theory brings out clearly the importance of the affective and emotional link between the child or adult and his environment. The no less controversial theory of sexuality indicates the extent to which we are controlled by our internal programming and how the dismissed affects that have not been worked through can play a negative part in provoking conflicts. Freudian theory has highlighted the role of aggressivity, and the first ethologists, like Konrad Lorenz, largely drew on that tradition. Freud's statements have proved invaluable in all manner of fields and have inspired numerous researches. For the rest, the richest areas are probably those in which different methods are mutually confirmed through their convergence: for example, those that compare psychoanalytic constructions emanating from adult analyses with direct infant observations (Bowlby, 1979, p. 4n).

Research concerning therapeutic *results*—in other words, the extent to which psychoanalysis and other therapies are likely to produce changes—assumes a particular importance. It is enough to refer to the works of Annemarie Dührssen (1972) and the more recent ones of Smith, Glass, and Miller (1980). All these studies show that psychoanalysis actually has the capacity to implement modifications, even if it is still very difficult to establish the precise values of them. In this respect, behaviour therapies have the advantage that they address only *a single* symptom, and any change is, therefore, easier to see. Psychoanalysis, aiming at different modifications in the representational world, in affectivity and the attitude to life, naturally offers a wider variety of aims, which are difficult both to define and to control. So-called "brief" psychotherapy provides a good model for establishing elementary, simple theses, permitting a clear and explorable over-all view, the effectiveness of which can be *compared* with other therapies—those, for instance, that are based on behavioural or cognitive *principles*. This type of research, which was instigated in the first place by Michael Balint and is marked by such well-known names as David Malan, Peter Sifneos, Habib Davanloo, Hans Strupp, etc., or institutions like the Menninger Clinic (Wallerstein, 1986), along with current work by Mardi Horowitz and his San Francisco group, all offer some hope of learning more on the subject.

If "effectiveness" is defined as the capacity of an intervention to implement a change, and "efficiency" as the effort employed in that intervention (in terms of energy, engagement, time), then, according to Smith et al. (1980), there is little doubt that psychoanalysis is *effective*: it brings about *changes*. So critics have transferred emphasis onto the problem of its *efficiency*: is it an economic undertaking? Could those same changes be achieved more simply, more quickly, sometimes more surely by virtue of their greater transparency and controllability? Indeed, although Freud's aim at the outset was the disappearance of symptoms through their being understood, little by little psychoanalysis has devoted itself to broader aims: the freeing from internal oppression, in fact the realization of a better life in terms of the individual's desires. It is certainly not easy to take account of such wide aims in reliable, credible, and comparable studies that can be reproduced. Yet psychoanalysis must be prepared to respond to these challenges. As I have said, for some searchers, the interventions inspired by analysis, the brief psychotherapies limited in time and circumscribed, are the ones that can lend themselves first and foremost to such studies. These treatments with limited aims have in fact proved competitive with other treatments not inspired by psychoanalysis—behaviourist treatment or hypnosis, for instance. Their effectiveness is thus proven. But the main question is the achievement of more general aims through longer treatments, and effectively we do not know whether the changes that are brought about present some measure of stability. This question remains open, and it is mandatory for intellectual integrity that it be considered as such for the time being.

Freud presented a vast synthesis encompassing research on such subject as the unconscious, sexuality, the dream, and the child, and he tried to create a bridge between them. The cultural environment, the philosophy of the unconscious, the new sexology, and the romantic interest in dream have indubitably provided material. Wedekind (1924) referred to the "century of the child", so an interest in our ontogenesis was also in the air. Nevertheless, the synthesis came from Freud, and his methodology encompassed what Kaufmann (1980) called "poetic" science. Already in 1896, Alfred Freiherr von Berger had said of Freud and Breuer: "They stand shoulder to shoulder with the poet." Furthermore, as I have shown, he invented a methodology, a work in mosaic with constructions and reconstructions of the theoretical scaffolding following Cromwell's motto, "You never get so high as when you don't know where are going" (Blanton, 1971, p. 32). His image of man is a synthesis of conscious and unconscious, of the transparent and the opaque, enabling an integration of the hidden and the conflictual. Obviously he has taken as his model of "man" the civilized adult in whom the hidden incomprehensible must be understood through heredity, through childhood residues, through the heritages of archaic epochs ["Ur-zeiten"]. In the end, in the programme of "Enlightenment", it is a cultural elaboration that largely corresponds to the ideals of the intelligentsia, not only of a particular period in time but of the whole twentieth century.

Now let us reflect on the less elementary areas of the theory and its proofs, on psychoanalysis as a cure, an experience, an interaction: on representations and the affectivity connected to them. It is certain that Freud's conceptual tool was insufficient for studying all aspects of psychoanalytic treatment, at least in our current perspective. There is no doubt that he has effectively been limited by his wish to proceed empirically according to the laws of the natural sciences. Already in his lifetime, different controversies flared up about treatment practice and what actually happens in it. The best-known of these is the one that brought him into conflict with his closest friend and collaborator, Sándor Ferenczi, from Budapest. Historically it is only today that we are able to grasp the importance of this confrontation and of others—with Otto Rank, for instance. We must bear in mind, however, that other schools of psychology were unable to compare such notions as "drives" and "meaning" before, paradoxically, we became familiar with process models through information theory and artificial intelligence. There is certainly scientific progress in this field; at the same time we are coming to the point of measuring how far psychoanalysis as a method, as an encounter, was

> . . . in the first resort an art of interpretation and set itself the task of carrying deeper the first of Breuer's great discoveries—namely, that neurotic symptoms are significant substitutes for other mental acts which have been omitted. It was now a question of regarding the material produced by the patients' associations as though it hinted at a hidden meaning and of discovering that meaning from it. [Freud, 1923a [1922], p. 239]

> At first the analysing physician could do no more than discover the unconscious material that was concealed from the patient, put it together, and, at the right moment, communicate it to him. Psychoanalysis was then first and foremost an art of interpreting. [Freud, 1920g, p. 18]

If, therefore, psychoanalysis first tried to understand the relationship between analyst and analysand in terms of *instinct* and instinctual energy, it subsequently added understanding through interaction, especially *interaction* by way of *language* and non-verbal behaviour (intentional or non-intentional communications, such as slips). Today psychoanalysis is presented as an "auto-reflexive" movement (Habermas, 1985) or a scenario from the past, and as the *understanding* of what is presented—that is to say, *expressed*. Sándor Ferenczi, Michael Balint, and Donald Winnicott have done a great deal to develop the understanding of man's communicative behaviour and to bring attention to bear on what could be called the dynamic of mental processing. One of the earliest attempts at information modelling took place in Zürich under the direction of Ulrich Moser (1969).

To summarize, Freud's relationship with science is a history of successes and failures: a resounding *success* in his ideological or cultural influence on man's way of thinking in the twentieth century in our culture; in the understanding of important individual psychological facts, such as repression or the importance of attachment and the development of bonds in childhood; definitely in its indirect influences on anthropologists such as Ruth Benedikt, Margaret Mead, and Claude Lévi-Strauss, on ethology from Konrad Lorenz to Norbert Bischof, on psychology and, especially, Piaget's method.

The acuity and animosity of critics can probably only be understood by confronting them with the *promises* implicit in psychoanalysis. This has often been understood as being able, in the name of science, to pronounce a once-and-for-all and final truth about human life and even happiness: an "infallible" message. Such hopes and expectations have been entertained not only by a wide public, but even by several of Sigmund Freud's followers. The methods of psychoanalysis should be able to *eliminate all* individual *unhappiness* from the world. Before and after the Second World War, it was even hoped that a better psychoanalytic understanding of aggression could put an end to wars and other undesirable sociable phenomena. As might be expected, this was not the case. Analysis can eventually clarify aspects of the human personality, especially where motivation and intellectual and affective activity is directed towards such a demanding enterprise—a sort of self-unmasking, a school of ruthless honesty without make-up—but obviously disappointment, sometimes intense, was inevitable.

Freudianism and Marxism, too, while having their roots in antiquity—notably Plato—are sustained by two traditions

simultaneously: *romanticism* and *rationalism.* Romanticism
provides the impetus, rationalism the attraction for the factual.
Both present themselves with the idea that we shall use a reliable method of thought and action that is likely to lead to a
future without any major dissatisfaction and significantly de-
creased conflicts. In short, they imply the belief that we shall
become what man "actually" is, or, as Kolakowski (1978) said
about Marxism, what his "true nature" consists of, as opposed
to his contingent empirical nature. This *utopian* thinking is
present in Freudianism, as in all the great systems of thought
of that era.

Furthermore, these expectations are not only of *happiness,*
but of *salvation.* In other words, psychoanalysis was carried
along by a movement the characteristics of which were utopian,
with a religious undertone. The utopian aim is perfect happiness. By the nature of things, such an aim is not easy to define;
its *criteria* can only be vague and its realization uncertain.
There ensue from it phenomena that are not readily compatible
with a scientific attitude: the aura, for instance, surrounding
people felt to be in possession of knowledge, or faith in the fact
that others could equally aspire to it, provided they strictly
follow all the directives. This has engendered an atmosphere
similar to other utopian, religious, or political movements. It is
true that psychoanalysis itself has considered such expectations, vis à vis the analyst, for example, as analysable material
that should be analysed. Jacques Lacan (1973) presents the
analyst as one who is "supposed to know", as someone the
subject regards as having more knowledge. But this is a dimension that should be analysed because, in actual fact, only the
subject knows. . . . However, these enlightenments do not get
the better of irrational forces creating and recreating gurus and
masters, like Jacques Lacan.

Those who believe that psychoanalysis can put everything in
order can only be disappointed by its partial successes, its
partial "insights". Here, as elsewhere, science is a method of
"trial and error". The great enthusiasm turned into enormous
disappointment, particularly in the United States, where, more
than in Europe, psychoanalysis was strongly idealized in the
1930s to 1950s. In Europe it was an object of criticism at first,
because it seemed to question values until then considered

traditional—religion, for instance, and sexual behaviour—and then for apparently not challenging these same values radically enough. Gradually, it should be appreciated for what it actually is: like all other sciences, a great project consisting of some major ideas and a modest contribution towards a better knowledge of man's internal world. This knowledge always calls for new research and new comparisons with other sciences in a development that is continual and never-ending—interminable.

CHAPTER EIGHT

Freud, psychoanalysis and its crucible

> *Nothing of him that doth fade*
> *But doth suffer a sea-change*
> *Into something rich and strange.*
>
> Shakespeare, *The Tempest*, Act I, scene 2

P sychoanalysis tries to be a discourse on the inexpressible, its intention being to grasp the non-intentional content of the word, "to explore" what is beyond volition and escapes the individual's control—the unconscious, which goes beyond immediate awareness and intention.

Is it chance that it bloomed in the Austro-Hungarian empire? Does it carry the marks of the cultural melting-pot in which it was formed and developed?

First of all: who gave birth to it? Do we find among its ascendants, as Freud intended, the thinker in maieutics, the writers who staged the "human passions" through the exceptional behaviour of their heroes, from Oedipus Rex to Hamlet, and the tragic consequences of them? Do we number among them philosophers and thinkers—the English empiricists and

German physiologists who built the scaffolding for theories constructing a scientific discourse? Do we find Goethe, the great recluse from Weimar, then Schopenhauer and Nietzsche?

It must be remembered that because of the unpleasant "truths" it expresses, the psychoanalytic discourse has given displeasure and put up "resistances". Doubt has been cast on its "scientific" status. Yet, if the psychoanalytic method is considered as a discourse about the discourse, as a mirror held up to man, all these problems become of secondary importance. It does not purport to proclaim the Truth, but "to each his own truth": a human effort to penetrate the secret that is hidden, concealed by taboos.

It has adopted as its motto Virgil's "*acheronta movebo*", until then personified in the image of the visionary poet of the romantic era—German or otherwise—then by Nietzsche and Rimbaud, those on the fringes, excluded, who went through "Inferno" and bore witness to their experience. Freud wanted to express himself as a scientist, not as a Shakespearean dramatist or a philosopher. Would he be able to realize the programme of Nietzsche's "Ecce homo" (1908)? This man—this "superman"—who was to understand his own and others' motivations, who was to integrate Apollo and Dionysius present in the myth of the Analysed Man, is in any case the spiritual grandson of Nietzsche.

The search for links, for relationships between psychoanalysis and the context in which it was born, inevitably leads to a recapitulation of the cultural history of Vienna at the end of the last century—a period that is viewed with increasing nostalgia (see Ellenberger, 1970; Erdheim, 1982; Jaccard, 1982; Janik & Toulmin, 1973; Kaufmann, 1980; Nielsen, 1982; Wohl, 1979, etc.). A study of the relationship between this melting pot and the individual creator—Freud—also risks embarking on a discussion of questions of influence and priority. Did psychoanalysis arise, fully armed, from the head of Freud, like Athena from Zeus', or is it the reflection of something that was "in the air", and Freud only had to reveal it? Clearly, here as elsewhere, the creator forges new connections, which make it possible, then, to integrate a "stream" into its culture.

Freudianism was born alongside *sexology*: if one remembers that sexuality was one of the most important taboos of bourgeois Victorian society, one realizes that the scientific habit facilitated the emergence of a discourse both respectable and fringe. The field of so-called sexology is confined to the "anomalies" of sexuality, while psychoanalysis tries to integrate an understanding of the "tenebrous depths of the heart", no matter in whom its hidden aspects exist, in order to follow the avatars and metamorphoses of representations in the "pure forms" of perversion, in daily fantasizing or sublimated in art and religion—in mysticism for example. It results in the creation of a homogeneous field in the sense of Terence's "*nihil humanum alienum esse puto*".

The creation of a scientific discourse on man which collated everything that can be known about him, including his hidden—unconscious—aspects, and the development of models of his "functioning", of his *representations*, of their ("economic") equilibrium and *links* with biology (the instincts)—that was the ambition of psychoanalysis as Freud conceived it and as it was set up in the extraordinary cultural dawn of the Vienna of the end of the nineteenth century.

If it is true, as Whyte (1960) has shown, that Freud did not "invent" the unconscious, it is clear that the cultural environment bequeathed him a problem that he, in turn, took up in an original way. Its originality does not lie in the fact that he approaches the unconscious, but in the way he does it.

The unconscious has an extensive historical background. Thus, for example, in a book by Carus (1848), the concept of the unconscious appears clearly from the start with consciousness emerging from it, and also the notion of the unconscious becoming conscious. This idea reached its apogee in the celebrated work of Edward von Hartmann (1869), *The Philosophy of the Unconscious*, which was a best-seller in the salons of the nineteenth century. Again, at the International Psychology Congress in Munich in 1897, Theodor Lipps declared that the problem of the unconscious is *the* problem of psychology, and in 1902 William James called the discovery of the unconscious "the most important progress in psychology" (Pongratz, 1967). The word "*unbewusst*" had already existed in Germany for a long time. Goethe used it, and it was widely employed by the romantics. In his "Poétique, ou introduction à l'esthétique" [Poetics, or an Introduction to Aesthetics], Jean Paul wrote in

1804 that the most powerful force in the poet was just that, the unconscious. In English the word "unconscious" was already used in 1751, by the Scottish philosopher Lord Henry Home Kames. In Geneva, Henri Frédéric Amiel showed remarkable introspection in his diary where he used both the adjective and the noun "unconscious". The word also figures in the *Dictionary of the French Academy* from 1878 onwards.

The historical background may locate Freud's spiritual ancestors; but it says nothing about the problems raised by his use of the concept of the unconscious. According to Freud, the word instigates a *process* aimed at widening man's awareness of his internal world. Every cultural work implies such an expansion of awareness. Sophocles, Shakespeare, and Saint Theresa of Avila already worked in this sense. Freud can be classed amongst them also, in the same way that he is close to thinkers like Schopenhauer and Nietzsche, writers like Schnitzler and Karl Kraus, and musicians like Gustav Mahler, one of his patients. With Sophocles he shares the meaning and direction of the intellectual process; he comes close to those in the second group in points of condensation and conceptualization, and in the way his thinking is elaborated, grasped, and named. Nietzsche's *id*, for example, becomes, by way of Groddeck, the Freudian id. (Nietzsche, 1886, says that a thought presents itself when "it" wants and not when "I" want, so that it is *falsifying* reality to say that the subject "I" is the condition for the predicate "think". It [*es*] thinks but so that "it" [*es*] is precisely the old and well-known "I" expressing our moderation, our hypothesis, our assertion, but not an immediate certainty. Finally, this "it thinks" is already too much: the "it" already contains an *interpretation* of the process and does not belong to the process itself.) In a different context, this element changes by being defined in terms of the drives—but it is clear that its affiliation remains within a certain continuity: thus *changing* and *continuity* of meaning are notable features.

The kaleidoscopic play of resemblances and continuity becomes a mosaic in the flux of the history of ideas. In addition, Freud's position on the fringe of the society of his day (Leupold-Löwenthal, 1980), his not identifying with certain aspects of the social and private life of his immediate environment, enabled him to wonder and to question. By thus detaching himself from the "here and now" to achieve a more global vision, he produced a dynamic description—today it would be called transcultural—of human functioning, relating it, at one and the same time, to the new ethnology, the culture of antiquity, and the great works of drama and poetry.

In a field like psychoanalysis, which endeavours to grasp and to conceptualize intrasubjective experience in order to construct a model of psychological functioning that accounts for unconscious processes, it is obvious that only culturally available tools can be used. Besides, they are the same tools that instigated the exploration.

To appreciate the influence of the environment on psychoanalysis, we might compare the links between Vienna and Freud, Budapest and Ferenczi, with, for instance, the more recent ones between Lacan and the Paris surrealists. *Furthermore*, in relation to the word "unconscious", "*si duo dicunt idem, non est idem*" [if two people say the same thing, it is not the same thing].

Psychoanalysis seeks to apprehend the "rationale", the logical order hidden in the human heart, the secret forces governing individual behaviour. Its development has been subject to the influence of German romanticism. Indeed, its vision of man owes more to the theme of the "*Doppelgänger*", to the destructive passions, than to concepts of associationist psychology, psychophysiology, and neurology, or to Nietzsche's id and Brentano's ego.

A closer study of the entrenchment of psychoanalysis as a cultural movement in its surrounding milieu necessitates following the development of the notion of *conflict*. All Freud's commentators acknowledge this to be a pivotal concept in psychoanalytic theory (cf. Laplanche & Pontalis, 1967, p. 359). It can be seen in the development of Freud's thinking on this subject that from the time of his *Studies on Hysteria* (1895d), as soon as he comes close to pathogenic memories in treatment, he is up against a growing resistance, which he interprets as the actual expression of a defensive intrasubjective battle against incompatible representations ["*unverträglich*"]. He regards these defensive activities as a major mechanism, first in the aetiology of the neuroses and then, more generally, in human life—since Freud would say that between normality and neurosis he finds only quantitative differences (1940a [1938], p. 184), and that the distinction is solely a practical one (p. 195).

It is true that the notion of conflict existed before Freud, but as a concept it was applied to open confrontations (between people, groups, or institutions) in military strategy or social problems. The classical French authors (Corneille, in particular), for their part, represented the moral combat between passion and duty. In Kant and Goethe we already find moral conflict. Freud made use of the idea to explain situations in which the internal combat *is not apparent* because of the existence of active underlying forces. So he uses the allegory of a battle between different psychic tendencies on the "internal stage", permitting a new representation of the phenomena.

Regarding "internal stages", let us look at Arthur Schnitzler's work. His themes plunge us into the Freudian world—so much so that Ellenberger (1970) could write that Freud's *Dora* seems to come out of the pages of one of Schnitzler's novels. On the occasion of the sixtieth anniversary of this man who had been a neuropsychiatrist before becoming one of the greatest Austrian writers, Freud confided that he felt he had avoided Schnitzler through fear of meeting his double ["*Doppelgängerscheu*"] (Henry Schnitzler, 1955). Like Freud, Arthur Schnitzler came from a family of enlightened Jews, detached from religious orthodoxy except for the traditions of his ancestors. He was born in Vienna on 15 May 1862, six years after Freud. His father, a professor at the University there, was the laryngologist of the Viennese upper middle classes and the operatic singers. Like Freud, Schnitzler studied medicine in Vienna and obtained his doctorate three years after him. Like him, he worked at the Vienna General Hospital, where he was also a pupil of Meynert. He was interested in hysterical loss of the voice, which was the subject of his first scientific publication in 1889 (his father's speciality—but treated by hypnosis). He made his first compromise between his talent for writing and his medical career by working in medical journalism for the prestigious *Wiener Medizinische Presse*. It was perhaps he who reported, in an article signed S., the famous session in which Freud talked about masculine hysteria [*Wiener Medizinische Presse*, 27 (1886): 1407–1409]. He wrote numerous accounts of medical works in the *Internationale Klinische Rundschau*, in which he commented on Freud's translations of Charcot and Bernheim. When Freud gave his celebrated presentation to the "Doktorenkollegium" on 4 October 1895, Schnitzler was the reporter [*Wiener Medizinische Klinische Rundschau*, 9 (1895): 662 et seq.]. Gradually he devoted himself exclusively to literature and the theatre. The "Anatole" cycle, the story of a young high-liver, confirmed his reputation. He gathered around himself a group of young poets and dramatists who took the name "Jung Wien" [Young Vienna]. His 1989 play, "Die Frage an das Schicksal" [The question asked of Fate]—from the Anatole cycle—presents a young man who hypnotizes his beautiful mistress to find out whether she is faithful. What is revealed is that her age is 19 and not 21. Anatole, alarmed, dares not pursue this further and wakens her. It is typical of the manners of the time to consider women as rather frail sexual objects. According to Freud, their superegos are less strong than those of men's. He also portrays the contemporary fear of what might be revealed of the unspoken, of what it is better not to know—namely, the fear of the unconscious. We are reminded of Ferenczi analysing Elma in order to know her feelings, and of Jung looking into the intimate life of his wife, Emma, in order to ameliorate their life as a couple.

One of Schnitzler's later plays—"Paracelsus" (1898)—took hypnosis as its theme. The hero hypnotizes Justina, who reveals certain things. No one knows how much truth there is in what she says, nor at what point she is wakened from her hypnotic state. Again there is the knowing and not knowing, the continuity of unconscious and conscious experience. In the play, Paracelsus remarks that memory is almost as deceptive as hope. This difficulty in grasping the truth gives us some indication of Freud's over-determination. Schnitzler is one of the first to write a novel—"Lieutenant Gustel"—in a "stream of consciousness" style, and this in 1901, as a parallel to Freud's free association, well before Proust or the surrealists. In this particular novel, the characters have dreams in the course of which recent events, memories from the past, and the preoccupations of the moment become distorted and confused in a variety of ways. Is this "Freudianism"? It seems to me that what we see here is the influence of the same cultural melting pot, in that fascination for the "decadent" interplay of certainty and uncertainty, for repressed sexuality, in the idealization of prostitution as freedom, at the same time as it reflects human misery, in the questioning of consciousness and in the interest in dreams. This is *fin de siècle* Vienna and the crucible in which scientific discourse was formed by Doctor Sigmund Freud, who uses all these elements like stones to construct a new anthropology, a new human science. According to Ellenberger (1970), the younger generation of Austrians saw in Schnitzler, after the First World War, "the prototype of the corrupt spirit of the decadent monarchy", "the frivolous life of the leisured class of Vienna". We have since realized that that "decadent" period of the monarchy was a time of questioning what turned out to be fundamental for the twentieth century. When Schnitzler, in "Ueber Krieg und Frieden" [On War and Peace], says that neither soldiers, officers, or diplomats, nor men of state really hate the enemy and that the true causes of war rest elsewhere, he is questioning myths as Freud does and forcing us to look at other possibilities without necessarily developing them (although he did evoke public opinion, the press, the perfidious intentions of a small number of individuals, etc.). But in this way, he showed men's incapacity to acquire an image of war. If one compares this work with Freud's "Thoughts for the Time on War and Death" (1915b), one is aware of the original way Freud went more deeply into subjects that were "in the air" at the time.

The play that Schnitzler wrote in the throes of physical suffering caused by otoscloris at the end of his life is entitled "Flucht in die Finsternis" [The Call of the Deep] (1917). It describes the subjective experience of a schizophrenic and the development of the illness leading to the murder of his brother, a doctor. It is quite clear that the two men were exploring the same area. Freud's aim was to create in science the intentions and psychological knowledge revealed by the great writers (Ellenberger, 1970). That is the reason he did not become either a Schnitzler or a Shakespeare, but someone else, at the same time remaining related to Shakespeare and a close contemporary of Schnitzler.

Schnitzler also describes the importance of antisemitism, which, he said, will never be estimated at its true value. He considered that posterity would scarcely be more capable of forming a true idea of the psychic meaning—almost more than the political or social meaning—of the so-called "Jewish question" at the time he was writing. For it was not possible—especially for a Jew in an official position—to ignore the fact that he was a Jew, since others did not do so, neither Christians, nor—and even less so—the Jews themselves. One had the choice of being insensitive, importunate, and impertinent, or sensitive, timid, and persecuted (Schnitzler, 1968/1981). He does go on to say, however, that after the apogee of liberalism, antisemitism existed, as it always did, as an emotion in many who were disposed to it, and as an idea highly liable to development; but it did not play an important role politically or socially. Even the word did not exist, and those who were unduly ill-intentioned towards Jews were scornfully called "Jew-eaters".

A great deal more could be said about the links between the *literary* life of the period and Freud's work. I shall content myself with quoting Wedekind [1864–1918], whose *Frühlingser-wachen* [Awakening of Spring] (1891) presents the stormy age of puberty, the mysterious and alarming awakening of the drives in the face of the dismal incomprehension and unyielding stupidity of the adult world. If one had to give a Freudian clinical illustration of adolescence, of the battle between the id and the superego, a better one could not be found. Furthermore, one sees the emergence of fantasies, with the appearance, in the cemetery, of phantoms both of a masked man and, in Loulou's attic, of an idiot, a black prince, a professor, and Jack the Ripper, the man who killed Loulou. . . . As in an analytic treatment, discourse and reflection are put in the mouth of the homosexual woman, the prostitute, murderers, tight-rope walkers. In other plays, Wedekind talks about the fatality of instinctual attraction (*Erdgeist* [Spirit of the Earth], 1895; *Die Büchse der Pandora* [Pandora's Box], 1904) and of prostitution (*Danse macabre*, 1905).

Another example is Karl Kraus, who in his diary *Die Fackel* [The Torch] was in the habit of reproducing amusing misprints, showing that the typography had involuntarily guessed, sometimes betrayed, the author's thoughts. Here again, the meaning of parapraxes is known to an important contemporary of Freud. But Freud goes further in linking the phenomenon to the problem of "wishing", of desire, especially as it appeared in dreams. This "wish" [*Wunsch*] sometimes assumed sexual dimensions, as Schnitzler, Wedekind, sometimes Kraus, and others were aware. Otto Suyka (1905), for instance, in his criticism of Forel's *The Sexual Question* and Freud's *Three Essays* (1905d), made sarcastic comments on Forel's work, whereas he put Freud's on the same plane as Schopenhauer's *Metamorphosis of Love*. For the latter, love life is still an object of metaphysical investigation, while in Freud it becomes an element of scientific discourse. It was Wittels who familiarized Kraus with Sigmund Freud's work. The accord and mutual esteem between Freud and Kraus eventually deteriorated, in part because of this same Wittels, who caricatured Kraus in the "*Mittwochs-Gesellschaft*" [Wednesday Club] (Nunberg & Federn, 1967) in Freud's presence and ended by writing a roman à clef about him entitled *Ezechiel der Zugereiste* [Ezekiel the Alien] (1910). This did not, however, prevent what might be called the "Middle-European spirit" being found equally in the works of the satirical critic and the humanist scholar. In the final analysis, both were interested in the same phenomenon: language—the one as a satirical polemicist, the other as a doctor who, even before inventing psychoanalysis, had published a book on aphasia. Freud does not hide his admiration. He is quoted by Szasz (1977) as saying that a reader who is truly a companion wants to compliment one's intuition, courage, and capacity to see what of significance is hidden in all that is insignificant (Freud, 1904, quoted by Szasz, 1977). And Kraus does not hesitate to use Freud's authority to protest against anti-homosexual legislation proposed in 1905. For his part, Freud issued a statement in Kraus' monthly *Die Fackel* [The Torch] on the complicated affair of plagiarism, involving W. Fliess, Svoboda, and Otto Weininger.

A country as full of contradictions as the Austro-Hungarian empire was sure to create greater sensitivity to the contrasts and multiplicity of forces acting upon the character—the nonunity of the subject was to be a theme in Robert Musil's *L'homme sans qualités* [The Man without Qualities] (1952).

In all that was written up to this time, decadence, uncertainty, and anxieties play an important role, but it should be realized that it is all behind a façade of security, as Schnitzler describes in his *Diary* (1968): "At the time, one believed that one knew what Truth, Goodness and Beauty were. There was a magnificent simplicity about life." In the midst of the modern period, systems seem to be stable, there is belief in the Truth—contrary to what is felt in our post-modern era, as Stephan Zweig recalls retrospectively and so powerfully in *The World of Yesterday: An Autobiography* (1947).

Freud showed that what we say does not always express our deepest feelings—that we seek, through our attitudes and appearances, to gain the approbation of others, and that the confusion and contradictions can be understood through the history of the individual. In order to state these facts, he had to have experienced Vienna. Only the history of this heterogeneous empire, unlike other states built up with some unity—a "Cartesian" coherence, for example—can contribute to the feeling of complexity in the life of this circle. The entrenchment of Freud's work in the atmosphere of the empire is also manifest, in his dream analyses, in the introduction of the notion of *censorship* (cf. Johnston, 1980). Freud no doubt had before his eyes the way the political writer functioned in that multinational conservative state, as he witnesses:

> If he presents them undisguised, the authorities will suppress his words. . . . A writer must beware of the *censorship*, and on its account he must *soften* and *distort* the expression of his opinion. . . . The stricter the censorship, the more far-reaching will be the disguise and the more ingenious too may be the means employed for putting the reader on the scent of the true meaning. [Freud, 1900a, pp. 142–143]

"The powerful" represent conscious control, and the little gnomes often jokingly distort the unconscious contents. Political conflicts provide an image of mental functioning. The following comment is clear (Freud, 1900a):

> We may therefore suppose that dreams are given their shape in individual human beings by the operation of two psychical forces (or we may describe them as currents or systems); and that one of these forces constructs the wish which is expressed by the dream, while the other exercises a censorship . . . [which] forcibly brings about a distortion in the expression of the wish.

In his memoirs, Ernst Lothar (1960) reports a remark of Freud's made after the carving-up of the Austro-Hungarian empire and his mother's death. Lothar had been wondering how one could continue to exist without the country one had lived for. Freud's reply was as follows: "The mother is the birthplace of each one of us. That we survive her is a biological fact since the mother dies before her children. . . . There always comes a time when an adult becomes an orphan." He referred to Lothar's comment that the country perhaps no longer existed. "Perhaps the country you are thinking of has never existed and you and I have been deluded. The need for illusion is also a biological fact. A time may come when you will realize that someone close to you is not what you thought" [translated for this edition from the French].

Later he says: "Like you, I was born in Moravia and like you nourished an enormous affection for Vienna and Austria although perhaps unlike you I knew what was going on beneath." It is a beautiful description of his so often evoked ambivalence. He knew the faults of a city that was the backdrop for the exploits of extraordinary minds. He was affectively attached to it while being very critical of it, but, as we have seen, his love for it prevailed. He wrote the following on a scrap of paper: "Austria-Hungary is no more. I could not live elsewhere. Emigration is out of the question for me." Only when the death instinct engulfed the country in all its cruelty and with the threats against his daughter Anna did Freud accept the idea that he might have to live elsewhere. So his work ran into a human tragedy, which was that of thousands and then of millions, some of whom, like him, succeeded in saving themselves, while others were the victims of the death instinct's total destruction. It must be admitted that Freud's allegedly pessimistic vision turned out to be rather prophetic. Sadly, he had to witness it himself.

I shall continue my theme with the words he wrote to Fliess on 5 November 1897: ". . . one still remains a child of one's time, even in what one deems one's very own"—including his scientific creation.

CHAPTER NINE

Construction of a theory

The cultural context of the Central European intelligentsia is grippingly described by Arthur Koestler in two of his auto-biographical books: *An Arrow in the Blue* (1952), and *The Invisible Writing* (1954). Koestler's quest—of an "-ism", whether socialism, communism, Zionism, or Freudianism—is typical of the whole intellectual atmosphere of his generation characterized by the search for new *keys* that will open up, much later than in France—with its Encyclopaedists, Voltaire, and the French Revolution—a secular understanding of man and the world. The hold of the Catholic church remained predominant in the nineteenth century. The Jewish intelligentsia were not liberated from the *"Stetl"*, small towns of their origins, and the traditional historical context, until this century (and then radically, and with some intellectuals coming from the traditional strata of the middle classes). It is a search for an explanation that will provide, in Haeckel's words, the solution to the *mysteries of the world* (1899). Just as Darwinism claims to solve the problems of the origin of man and the natural sciences to solve those of the material environment, so an explanatory model of the mind is sought after, and Freud considered he had found a *doctrine* that responded to it. It was for his authority to defend it, and for his circle to believe in it and to use it. It is in this *epistemological attitude* that the history of psychoanalysis must be understood and the consequences of it addressed. Freud had the firm conviction that it is a *science*—the last-born of the natural sciences—which had to be *defended* against every anti-enlightening, anti-scientific, conservative reaction. In that respect he shared the authoritarian attitude of most Central-European university teachers of the time while paradoxically savouring Heine's irony (Freud, 1900a, p. 490; 1933a, p. 161; letter to Jung, 25 February 1908, etc.):

> *With his night-caps and rags of gown*
> *He stops up the gaps of the universe.*
>
> [Heinrich Heine, in E. Jones, 1953, p. 214]

The present generation, which has experienced rapid changes in scientific information and models (in physics, for example), is relativistic and pragmatic. It doubts on principle the validity of authority, which perhaps causes difficulty in understanding the atmosphere in which psychoanalysis grew up and with which it has been surrounded until very recently.

Today there is "no scientific method", according to P. B. Medawar (1967); a scientist is first of all someone who "tells stories", within certain limits: he is obliged to verify their authenticity. So "having ideas is the supreme achievement for a scientist". The search for systems "trying to cover the whole truth", in which the scientist is idealized and looked upon as a Dr. Faustus, has been abandoned. We are resolutely in *post-modernity*, where there are only "islands of determinism".

A decline in universalist discourse and pseudo-metaphysical doctrines is the characteristic of our present epoch. It is not that of Freud's. He was a product of his time and did not share our lassitude towards "theory". Accounts of progress—particularly scientific progress—still captivated him. In Vienna, around 1900, "-isms" flourished: Freudianism, Zionism, neo-positivism (in the Vienna circle), and expressionism. They testified to a need for doctrines that have been influential right up to our present post-modernity. Even though he created Freudianism, Freud remained, paradoxically, faithful to his scientific credo: biology. He consistently had great confidence in the development of biological treatments: his experimentation with cocaine can be considered as in the field of psychopharmacology (Freud, 1856–1939, 1884e, 1884g, 1885a, 1885b, 1885e, 1887d). Stricken with cancer, he did not have any analysis, although both Ferenczi and Groddeck suggested it, but underwent a treatment then considered as endocrinological: a surgical intervention using Steinach's method (involving the ligature of the deferent vessels on both sides—Jones, 1957, pp. 111–112). He by no means thought that the capacities of psychoanalysis to induce changes, to treat people and to help them, prevailed over all other possible interventions, particularly biological ones. A literary psychoanalysis, detached from its biological substrate, was certainly not one of Sigmund Freud's prime intentions.

Marginality, legitimization

In Central Europe, psychoanalysts were, even among the intellectuals, rather marginal. With the exception of a few doctors (like Heinz Hartmann), many had no university degree (Anna Freud, Erik Erikson); either they acquired one later (Victor Tausk took a medical degree—see Alexander, Eisenstein, & Grotjahn, 1966, p. 235; Roazen, 1969, pp. 317ff.), or they were recognized little or not at all in their professional communities (like Ernst Kris and Géza Róheim). These were the "intellectuals of the left" in their era, considering themselves the inheritors of the Age of Enlightenment, acting in the name of progress, of the liberation of man and social justice, and trying to influence the development of their natural community. This was the case of the Budapest group—Edith Ludowyk-Gyömrői, Sándor Rado, René A. Spitz, etc.—and of those who were later identified with the "Freudian left"—Otto Fenichel, Wilhelm Reich, Muriel Gardiner, etc. (cf. Jacoby, 1983). The wider circles of society adopted or recognized psychoanalysis only much later and in these exceptional situations (e.g. in France in the 1950s and 1960s). It is fascinating to see how the first-generation analysts, who were all marginal, changed at a certain point in order to become respectable people practising an established profession, which was itself respectable. James Strachey, from the bohemian Bloomsbury Group, made Freud more "respectable" in translating him, introducing concepts like neutrality (instead of *"Indifferenz"*), and in some respects "correcting" Freud by incorporating Greco–Latin terms ("cathexis", "ego", "id", etc.) into the texts. Anna Freud, even without a degree from Vienna, had the manifest wish to direct a Clinic in London, which she later managed to do with the help of other rich American members of this Freudian establishment. Those who previously had met in cafés gradually occupied offices in universities and institutes endowed with "classes", places where the profession of psychoanalysis could be learned.

Some émigré analysts, especially in Britain, the United States, and, in some cases, Northern Europe (Holland, Sweden, and Germany), then tried to stand as *"experts"* of the *human mind*, in the manner of other intellectual experts. This situation is a delicate one, because of the lack of recognition of medically qualified psychoanalysts as doctors by their colleagues—especially under the influence of biology and information theories of mental processes. Analysts with psychological and literary backgrounds experience similar difficulties, though to a lesser extent. These are the people who believe in a critical "metadiscourse" from an *accepted marginal* position as the best way to maintain a clear situation, an originality of thinking, but at the cost of a certain isolation, which, in post-modern societies, does create some problems. Between the two kinds of knowledge—the one positivistic, linked to technique, the other critical, reflective or hermeneutic, which ponders over interhuman relationships in given societies—there is no doubt that psychoanalysis tends to belong to the latter. Whether, therefore, it is an "Arts subject" or a science of the mind in Dilthey's sense, as Jean-François Lyotard wonders, is another question. Like him, "we do not follow the solution of sharing" (1979).

According to Lyotard (1979),

> scientific knowledge is not the only knowledge. It has always been in competition, in conflict with another kind of knowledge which, for simplicity's sake, can be called narrative. . . . This is not to say that the latter might prevail over it, but its model is linked to ideas of internal equilibrium

and conviviality in comparison with which, contemporary scientific knowledge pales, especially if it has to be subject to externalization in relation to people who pretend to know and an even greater alienation from its users than before.

While presenting itself as a science (even if its legitimacy as such has remained a problem in view of the difficulties in verifying it), psychoanalysis has promised this kind of *narrative* knowledge linked to the idea of inner equilibrium and some notion of *pleasure*, a seductive discourse promising a certain happiness.

The failure of cheap "mini-analyses" following the explosion of psychoanalysis in the 1960s, mostly in French-speaking Europe, less so in Northern Europe and Great Britain, obviously diminishes its appeal to some extent.

Freud assumed that having created psychoanalysis, whoever followed his suggestions was likely to be able to practise it as he did. This legitimization through his *personal* authority has later been replaced by an *institutional* legitimization (the institution, i.e. the psychoanalytic movement, knows what an analysis is, and it is the legitimate guardian of the right to authorize psychoanalyses). This legitimization is interrupted by episodic returns to personal justifications by charismatic personalities (Melanie Klein, Jacques Lacan), to what is almost an addition: "If psychoanalysis is practised according to my rules, it is beneficial for the analysand". The analysand who "is authorized" to be analyst is only a striking phenomenon of a progressive development towards the *autonomization* that is being increasingly advanced by people, particularly in the liberal professions, in the evolution of our societies. But the historical change from authoritarian clinical practice to a more autonomous and more *responsible* style should not be evaded in the name of the permanence of the unconscious; this latter is, without any doubt, partly nourished by social and environmental factors that play an undeniable part in the internal life of subjects.

Historically, psychoanalysis pursues several aims. One of them, the constitution of an *autonomous* subject, by giving him the means to *emancipate* himself from *alienation* and repression, may be regarded as one of its constants. Its legitimization as a science, on the other hand, remains more problematic. Its methodology is the object of criticism, despite some convergence of its results with those coming from other sciences. Its success remains in part *ideological*—ideals of liberty and autonomy are its flag—and in part linked to its *pragmatic action* in individual treatments.

There are two possible viewpoints in the appreciation of the history of ideas in psychoanalysis. One takes Freud's work as having been accomplished, as a more or less closed system, which implies reservation vis-à-vis innovatory, "post-classical" contributions. The other considers it as a science in full development, far from being fossilized, and able therefore to acknowledge that different awarenesses may work together towards the discovery of new phenomena, new aspects, and new perspectives. Where are yesterday's reservations with regard to Winnicott? Even the passion of Bion's critics seems to have abated, and the critics of the "problem" Melanie Klein have practically disappeared. We should not forget that the psychoanalytic movement would undoubtedly be impoverished if, at the time, conservative "prudence" had had the upper hand. The same goes for Ferenczi, who, pushed away, almost absent from the collective memory of psychoanalysts and of teaching for decades, is today encountering a renewed interest: his contributions seem important and likely to open new pathways for rethinking certain fundamental problems. Ferenczi has certainly been the "gainer" in the controversy of which he was the object during the last years of his life, his ideas on transference and countertransference, empathy, and affective communication having profoundly influenced contemporary psychoanalytic practice.

Another personality who, although an integral part of psychoanalytic history, is insufficiently known in Europe is Franz Alexander. He was the son of Bernard Alexander, professor of philosophy at the University of Budapest and highly admired for his prodigious culture—a sort of "intellectual leader" of a whole generation of young people at the end of the century. Franz was himself an original thinker in psychoanalysis, unhesitatingly raising questions as "delicate" as the "attitude of the psychoanalyst", which puts him in line with his compatriot Sándor Ferenczi. Some evidence attributes to Freud attitudes similar to those Alexander advocates later. According to Natterson (1966):

. . . when, for instance, Reik had difficulty in eliciting transference feelings from a female patient, Freud's advice was
simple, "Make her jealous." So, on the next occasion, as
this patient was leaving, Reik warmly greeted the next
patient—also a woman—in the hearing of the problem
patient. During the following session, the previously aloof
patient became furious with Reik and freely expressed her
anger, and the analysis proceeded more effectively. [p. 256]

This episode, and other similar anecdotes, raise several problems: one concerns "manipulation" in psychoanalysis. For instance, Ferenczi writes about Jones that he (Jones) is afraid
that Ferenczi will talk about what he has heard in analysis, and
Ferenczi begs Freud not to mention their correspondence in
front of Mrs Jones (Fer. to Fr., 17 June 1913; see also Fer. to
Fr., 7, 17, and 23 June, and 5 August 1913; Fr. to Fer., 8 June
1913). It demonstrates the existence of a general *social climate*
present even in the analyst's consulting room. At the time, the
analyst would be considered advanced in his knowledge of
the unconscious and authorized "for the good of the analysand"
to impose or manipulate, though, obviously, no feeling of dis-
respect was shown towards patients. In the Freud–Ferenczi
correspondence, we come across all kinds of attitudes that
would be described as manipulative today. By his prejudices
and his "technique" the analyst participates in this same climate. The social habits that are generally acknowledged create
the atmosphere in which the relationships are formed. He can
only *detach* himself by his marginality and by his attitude of
questioning—but, obviously, *never perfectly.* As for Alexander's
proposals, one wonders where the line is drawn between
occasional manipulations and their systematic employment.
Clearly, Alexander's aim was to accelerate the progress of analytic treatment, or at least of the psychotherapies derived from
it. The problem of ever longer analyses is a very real one, and it
is certainly important to consider this subject, as Balint did.
That Alexander *dared to express* these facts is not appreciated
in the psychoanalytic movement, any more than that he turned
towards new fields that had been unapproachable until then—
like the psychosomatic illnesses. The search for dialogue with
other schools of thought in psychotherapy created animosity
for him. It was easy for Alexander's critics to quote Freud, who

did not wish his pupils to enter that field—probably out of concern to maintain a pure methodology. Thus, psychoanalysis runs the risk of becoming rigidified—and of distancing itself from innovative thinkers.

The fringe society of yesterday has become the keeper of the Grail today.

CHAPTER TEN

Freud, his disciples, and the psychoanalytic "movement"

Freud's relationship with his disciples can only be understood in the historical context of his culture—viz., that of a Viennese professor and his disciples in a patriarchal atmosphere in which benevolence, advice, and "privileges according to age" (cf. Freud's letter to Jung, 31 December 1911) all play a significant part; a hierarchical society if ever there was one, with the implicit assumption of a hierarchy of knowledge. Those in his circle were fascinated by his discoveries because they held the promise of opening horizons and going beyond personal and familial problems. It should be remembered that Freud analysed his daughter Anna twice, that Jones had his children analysed by Melanie Klein (and tried to hide the fact from Freud), and that Anna Freud analysed the children of Dorothy Burlingham, with whom she later shared a household.

Grafted on this group situation are various specific factors induced by Freud's personality and by his *fantasies*—or, at least, what we can guess about them through his writings. Claude Le Guen (1972) shows how Freud's fantasies, as they are expressed in *Totem and Taboo*, and his identification with the "father" and with "Moses" influence his attitude towards Jung. (He writes to Jung: ". . . if I am Moses, then you are Joshua and will take possession of the promised land of psychiatry, which I shall only be able to glimpse from afar"—letter to Jung, 17 January 1909.) Freud had a *complex* relationship with his disciples. He was the analyst of some of them and remained an idealized (as was the case with Ferenczi or Jones' mistress) or an unanalysed (Abraham, Rank) transference figure. He *expected* a great deal from them, assigning them specific *roles*: Jung's was, as a *"goy"* [a non-Jew], to assume visible leadership and be somewhat of an ensign [*"Aushängeschild"*]. Ferenczi's task was technique, etc. He was keen to retain his authority, as is demonstrated over an incident with Jung who was said to be very disappointed about certain facts (Jung, 1962).

At the same time, in keeping with the current fashion, he maintained an atmosphere of secret connivances, alliances, complicities—with Abraham against Jung, with Ferenczi, Jung and others against Jones, etc. [For example, in a letter to Abraham (1 June 1913), he writes: "Jung is crazy." In letters to Ferenczi (5 March and 8 April 1927), he talks about Jones' aspiration towards complete sovereignty in the psychoanalytic movement; he warns: "Be circumspect with Jones", in another letter to Ferenczi (9 June 1927), and Ferenczi writes (30 June 1927): "All broadmindedness seems to be missing in Jones". Freud also discusses Melanie Klein's influence on Jones and writes: "I do not want Jones to become president" (5 July 1927; Fer. to Fr., 13 July 1927, etc.].

With regard to this it is worth quoting two passages from Roazen's biography of Helene Deutsch: "Freud may have been a holy figure to Helene, but she had reservations about him as a therapist." In the first few sessions of her analysis "Abraham showed her a letter from Freud in which he instructed his disciple that this was a marriage (of Helene with Felix Deutsch) that ought not to be disrupted by analysis" (Roazen, 1985, p. 193).

In addition to this is the astonishing principle that Freud expresses through a quotation from Goethe: "After all, the best of what you know may not be told to boys" (Freud, 1900a, p. 142). Thus, for example, in suggesting that Ferenczi write a

work on countertransference (Fr. to Fer., 28 January 1912), he
specifies in a letter to Jung (31 December 1911) that it should
be circulated only among the initiated. Looked at more closely,
these relationships are much more complex and cannot be
qualified simply as "tolerant" or "intolerant".

The Master

According to François Roustang, Freud had no scruples in ex-
ploiting the relationships he entertained with his disciples, im-
plicating their independence. They were almost totally given
over to their master. He even went so far as to "plagiarize"
their ideas, appropriating what suited him. This was so with
Abraham's ideas on depression and melancholia. In a letter to
Abraham he wrote how invaluable the latter's observations
were and that he drew from them "unscrupulously" whatever
was useful for his own paper (Freud, 1965a, letter to Abraham,
4 May 1915). In some editions of *The Interpretation of Dreams*
(Freud, 1900a), Rank's contributions were included, only to be
thrown out subsequently, after Freud had quarrelled with him.
He took up Isidore Sadger's idea (Duruz, 1985) of narcissism, a
term introduced by Näcke (see Ellenberger, 1970), Nietzsche's
"id" via Groddeck (Freud, 1923b, p. 23), and Sabina Spielrein's
death instinct (1912). These were all ideas that, in different
forms, had already appeared in von Schubert, Novalis, Metch-
nikoff, Nietzsche, Fechner (Ellenberger, 1970), and, very early,
in Stekel as "Thanatos" (Jones, 1957; Roazen, 1971). In all
these exchanges of ideas, he was the master, and there was no
question as to "ownership". Even granted that in this communal thinking of the Vienna group at its inception, "mine" and
"yours" had very little, if any, meaning, Freud's relationships
with his students were no less difficult.

Paul Roazen (1971) and François Roustang (1976) bring two
different viewpoints to bear on this situation. In Roustang's
view, being a student of psychoanalysis implied an unresolved
transference relationship, with powerful affective forces, hopes,
and idealizations linking the student to the professor. On his
side, Freud, the Master, needed his disciples to ensure that his

theory remained a theory during its construction and did not become a system of either delusion or illusion. It is easy to see how relationships based on this kind of submission, to the point of "sacrificing one's own intellect", as in religious (the Jesuits, for example, in their "*Omnia ad majorem Dei gloriam*" [All things for the greater glory of God]) or even in fanatical systems (for a psychoanalytic study of fanaticism, see Haynal et al., 1980), may be abruptly transformed into their opposite in an *impassioned* search for *independence*, so assuming an attitude of system opposition for the sake of liberation. This was so for some of Freud's closest followers: Otto Rank, in part Sándor Ferenczi, with some limitations Carl Gustav Jung, and many others.

The Master needs his students because he is identified with the *Cause*: what is done for him is, in fact, done for the Cause. In a letter to Abraham (11 October 1908), Freud writes clearly that he does not demand sacrifices for himself, but for the Cause. There is an uninterrupted circulation from the Professor to his theory and then from his theory to the *institution* that he founded.

Freud played on his age and his paternal role. He writes to Jung about Ferenczi,

> . . . who thought me cold and reserved and complained bitterly about my lack of affection, but has since fully admitted that he was in the wrong and that my conduct had been well advised. I don't deny that I like to be right. All in all that is a sad privilege, since it is conferred by age. The trouble with you younger men seems to be a lack of understanding with your father complexes. [Freud to Jung, 31 December 1911]

There seemed to be a certainty in the ability to discern the motives of others; it was as if, by saying, "you have a father complex", the problems with his disciples would be resolved. Yet obviously this is not the case! Jung expressed the situation clearly when he said (in a letter to Freud, 3 December 1912) that psychoanalysts abuse the analysis if they use it "to devalue others and make insinuations about their complexes". Freud's reply was to the effect that we should pay greater attention to our own neuroses than to our neighbours' (Freud to Jung, 5 December 1912).

These affective interactions and enmeshed personal feelings lead to another, equally complex, problem—viz. the hopes and expectations of salvation that are put into the psychoanalytic movement, which have a certain religious quality about them. Jones has already referred to them, not without irony (the psychoanalytic "movement" as he called it, putting it in inverted commas in order "to pillory it so to speak"—1959, p. 205).

The "movement"/the Cause

For a student, as the member of a group, it is a major difficulty to maintain one's creative autonomy. It is a situation similar to that of theologians caught in a patriarchal authoritarian and dogmatic system (one can quote Hans Küng and other contemporaries, but there is also no lack of examples in history). Freud's legacy, *the psychoanalytic movement* (Jones, 1959, p. 205), included a secret Committee set up in 1912 in which Freud perhaps perceived a touch of puerile romanticism. This again, it should be emphasized, is characteristic of the age. (Franz Molnar, a popular author in the Austro-Hungarian Empire, perpetuates its memory in his novel *A pál-utcai fiuk* [The Boys of Pál Street], a story of children in the Budapest suburbs, which was highly popular in Central Europe at that time.) It must also be realized that, as Jones (1959) noted, this movement is not dissimilar to ideological, religious, political, even fanatical movements and organizations:

> . . . our would-be scientific activities . . . partook rather of the nature of a religious movement, and amusing parallels were drawn. Freud was of course the Pope of the new sect, if not a still higher Personnage to whom all owed obeisance; his writings were the sacred text, credence in which was obligatory without discussion on the supposed infallibilists who had undergone the necessary conversion, and there were not lacking the heretics who were expelled from the church. It was a pretty obvious caricature to make, but the minute element of truth in it was made to serve in place of the reality, which was far different. [p. 205]

Gellner (1988) tackled this aspect of the Movement as a sociologist in a work that is basically negative towards psychoanalysis, yet highlights some indisputable characteristics. But was the psychoanalytic movement not the reflection of an age that, in the name of science, considered itself to be very close to finding a definitive solution to human problems?

What does or does not belong to Freud's work is the decision of the group—a sort of collective relying on sometimes nonscientific criteria. So, as his work on Wilson was difficult to integrate into it, it found no place in the *Standard Edition*, nor did the study on aphasia (Freud, 1891b) or those on cocaine (Freud, 1884e, 1884g, 1885a, 1885b, 1885e, 1887d), etc.

The psychoanalytic movement was put into concrete form by the creation of the International Psycho-Analytical Association. One of its founders, Ferenczi, formulated an astonishing criticism which deserves to be quoted here (Ferenczi, 1911 [79]):

> I know the excrescences that grow from organized groups, and I am aware that in most political, social, and scientific organizations childish megalomania, vanity, admiration of empty formalities, blind obedience, or personal egoism prevail instead of quiet, honest work in the general interest. [p. 302]

In 1928, not without pride, he adds (Ferenczi, 1928 [306]):

> For eighteen years the International Psychoanalytic Association has continued on my initiative. . . . I believed, and I still believe, that a fruitful discussion is only possible amongst the holders of similar lines of thought. It would be as well for those with a basically different view-point to have their own centre of activity. [translated from the German]

Yet, in Roustang's view (1976), this organization resembles a church, and Le Guen (1972) thinks that in choosing the youngest of his "sons", Freud re-established the fantasied situation in *Totem and Taboo* of the danger of being killed by the youngest brother. As a result, he entrusted the organization to a woman—and not just any woman, but his *daughter*, which can be interpreted as an attempt to avoid the problem.

* * *

Why is it that the psychoanalytic movement generates a certain scorn, even antipathy, among scientists, particularly the epistemologists (from Popper to Grünbaum)? Some critics of Freud from that field—Zimmer (1986) for instance—consider that psychoanalytic works are so vague and that their authors adopt such diverse epistemological positions and such multiple interpretations and readings of Freud that proof of error ("falsification" in Popper's terms) cannot be brought. Indeed, how is one to choose between authors who consider Freud as a practitioner of hermeneutics and those who rely on empirical or even experimental confirmation to support some of his fundamental theses, such as infantile sexuality, repression, etc.? Is it not time, a century later, that we acknowledge that, in the first place, Freud was the launcher of ideas, the creator of a project and of a technique for exploring man? May some of his ideas enter science, under different headings, *after validation* according to current scientific standards. Freud viewed psychoanalysis as an art of interpretation ["*Deutungskunst*"]. From there he tried to construct a science using the concepts of defences, drives, infantile sexuality, development, etc. According to Ricoeur (1965), the dichotomy between "hermeneutics" and "energy" is indubitable. Stand-points as passionately anti-hermeneutic as Grünbaum's (1984) do, I think, leave aside one of the essential aspects of psychoanalysis. Seeing *only* interpretation, only "art" (the *art* of interpretation), gives short shrift to Freud's scientific ambitions, which were essentially to *contribute* to the creation of a *science*. What is probably most astonishing in Freud's work is the complexity with which he constructed the work, piece by piece, like a mosaic, along "*complementary*" lines (dynamics, energy, etc.), in a system that makes him one of the precursors of post-modernity. To see in it syntheses, dogmas, "treatises" of psychoanalysis is a mistake and a misconception of the nature of his work. Locating all of it as external to science would be as unfair as claiming that *all* of its construction is *ipso facto* a science. In its entirety it is rather a fresco, a tableau in which certain elements can be verified from the viewpoint of the patient construction of a science in dialogue with others, by examining the character, appropriate or otherwise, of methodologies that are offered to this sort of construction today.

Similar phenomena are repeated in relation to other charismatic leaders like Melanie Klein and Jacques Lacan (Roustang, 1976), which is proof that they are linked to expectations of safety, to the need for certainty and for emotional security in man.

Why—like Jones (1955, 1957, 1959)—call on motives and give interpretations to explain the deviation in some of Freud's followers, and not to explain the "loyalty" of others? Explanations and interpretations, often very obvious ones, can be found as easily for the one as for the other; in Vienna for example, remaining loyal to Freud could be more lucrative in providing an assured income, since Freud passed on to his colleagues patients who had been referred to him. I am not suggesting that it was material interest that kept the analysts in Freud's circle, but, rather, that reasons for both "departures" and "faithfulness" are complex and have elements of the material as well as the sublime in them.

Reflections on the history of psychoanalysis ought to bring about its *demystification*, which is the very aim of psychoanalysis as a form of "enlightenment". In upholding the illusion of it as a salvation, I would say that the consideration and recognition of its real evolution would certainly have a liberating effect on its practitioners, for their responsibility in recreating afresh the analysis of every analysand cannot be nourished by omnipotence, by the omniscience of another person, nor by the belief that the "system" of Psychoanalysis with a capital P has solved all the problems in advance. That would be the incarnation of wishful thinking and wish-fulfilment—of those unrealistic wishes announced (or denounced) by Freud.

There are limits to the ability of historiography to highlight the master–disciple relationships of authority between Freud and his followers. Such limits are imposed by human temporal and cultural considerations. But despite these limits they should help us, as psychoanalysts, to connect with our more fraternal age, which enables each one of us to assume personal responsibility without systems that are too authoritarian or protective. When Freud talks about secular pastoral work ["*weltliche Seelsorge*"] (Freud, 1927a, p. 255), when he is preoccupied with the anticlerical battle of the time, he is a true son of that Vienna which, having escaped the French Revolution, was not to feel the benefit of the new winds of enlightenment until towards the end of the nineteenth century. That *fin de siècle* is the same as central Europe's, where religion was beginning to yield to lay thinking and to transformation into "-isms"—Zionism, Austrian Marxism, Nazism, not to mention the various "-isms" in art and literature: surrealism, expressionism—all of which could present systems of belief likely to replace lost faith. . . . At the end of this century those beliefs and ideologies are, in turn, collapsing. Psychoanalysis ought to profit from this and submit itself to critical thinking in the same historical perspective as psychoanalytic questioning.

Freud's complex and difficult relationships with those around him—clinicians who considered themselves quite simply as "his pupils" and who, for the greater part, already had a good deal of *professional experience* (like Ferenczi and Jung) and a knowledge of areas with which Freud himself had little acquaintance—gave way to the institutionalization of psychoanalysis. The original fringe group increased gradually with people who, by the nature of things, desired more power, which for them implied greater respectability. Strachey's translations, with their classicizing of some of Freud's words and changing others that were deemed insufficiently "scientific" for his taste, are part of this trend, in the same way as is the acculturating of psychoanalysis in the United States and the exclusion of non-medical practitioners there. The small fringe group formed into a society, a profession was organized. And so, more and more, we come to the *professionalization* of psychoanalysis, and probably to the perspective of compulsory membership of the medical profession in the United States.

Dissidences?

Dissidence refers to a historical movement in which a majority opinion is assumed to be in possession of the truth and where a minority, because of its dissidence, finds itself—"through its own fault"—thrown *out* of the group. "Dissident" is defined, according to the *Oxford Dictionary,* as "disagreeing, esp. with an established government, system, etc.". Obviously it is an ex-

pression from the vocabulary of political totalitarian or religious, indeed pseudo-religious (messianic), movements, which
consider it necessary, for their action to be effective, that people
are loyal to, and have solidarity with, all the details of a political
programme. By contrast, movements based on the Voltairean
ideal and therefore liberal, indeed democratic, are founded on
the principle of free speech, in the hope of finding a solution that
takes account of both majority *and* minority opinions. *Loyalties*
are defined minimally through elements considered as fundamental to the very coherence of movements and dialectics capable of ensuring an exchange of ideas and actions. However, it is
not postulated that there is only one truth, that opinion should
be uniform, and that all aspects of life ought to be regulated by
one system of thinking (hence the term, "totalitarianism").

Some of the most important figures, like Ernest Jones
(Brome, 1982; Paskauskas, 1988; Steiner, 1985), and surpris-
ingly, James Strachey, begin to emerge from the historiography
of these latter decades. They undoubtedly represent an anxious
tendency to obtain some respectability for psychoanalysis, particularly at the time of its entrance into the Anglo-Saxon world.
It is a psychoanalysis probably closer to a medical conception
like Ferenczi's than to Freud's or others'. Michael Balint (letter
to Ernest Jones, 31 May 1957—cf. textual heading for chapter fourteen) thought that the study of these "cases"—and he
singles out two in particular: Ferenczi and Rank—would be
instructive to history. Those analysts who were preoccupied
with technique—people like Stekel, Ferenczi, and Rank—provoked some disquiet in Freud, who was sceptical about any
attempt to produce a more effective cure and who regarded
psychoanalysis above all as an instrument for understanding
man. As he grew older, it is a well-known fact that he became
more and more pessimistic about the efficacy of his discovery
(Freud, 1926e, p. 220; 1937c, pp. 224, 247).

For Ferenczi and Rank (1924) to have emphasized the role
of repetition for the new relationship is one thing; that over and
above reliving the Oedipal situation they carefully analysed
conflicts arising out of the relationship with the mother is quite
another matter. The choice of certain scientific themes seemed
to upset Freud. Ferenczi, as well as Rank and Jung, explored
the world of maternal relationships. After a period of active

technique (Ferenczi, 1919 [210], 1919 [216], 1921 [234]) of encouragement and other directives, Ferenczi arrived at the peaceful, unimpassioned atmosphere of the giving maternal attitude, which he emphasizes and which should create the conditions for the presentation and possible working-through of early traumata (Ferenczi, 1929 [287], 1931 [292]). It is reminiscent of Winnicott's "holding" technique. With Rank (1924, 1926), when the pre-Oedipal maternal transference acquired major importance, it meant that he extended Freud's ideas and rooted primary narcissism in the intra-uterine state. This should facilitate a new departure on the basis of those narcissistic sources which, according to Lou Andréas-Salomé (1966), nourish creation. As we know, Freud was positive at first. Unlike Abraham and Jones, he was considered to be balanced and careful. About Ferenczi and Rank's *The Development of Psychoanalysis* (1924 [264]) he wrote to Abraham as follows (15 February 1924):

> I value the joint book as a corrective of my view of the role of repetition or acting out in analysis. I used to be apprehensive of it, and regarded these incidents, or experiences as you now call them, as undesirable failures. Rank and Ferenczi now draw attention to the inevitability of these experiences and the possibility of taking useful advantage of them. Otherwise the book can be regarded as a refreshing intervention that may possibly precipitate changes in our present analytic habits. [Freud, 1965a, p. 346]

In another letter to Abraham, Freud writes about Rank's book on birth trauma (4 March 1924):

> Let us assume the most extreme case, and suppose that F and R came right out with the view that we were wrong to stop at the Oedipus complex, and that the really decisive factor was the birth trauma, and that those who did not overcome this later broke down also on the Oedipus complex. . . . Further, a number of analysts would make certain modifications in technique on the basis of this theory. What further evils would ensue? We could remain under the same roof with the greatest equanimity, and after a few years' work it would become plain whether one side had exaggerated a useful finding or the other had underrated it.

From all accounts, it seems that Freud *oscillated* in his appreciation of the problems of *pre-oedipal* versus *oedipal*, as he did later over the acceptance and refutation of Rank's theses. Reading these documents, it is difficult to resist the impression that what became a *mythical* rupture was, in fact, a dialogue that was pursued until it gradually died out. In the end, Rank considered it better to distance himself geographically from Freud and to live first in Paris and then in the United States; it did not give the impression of a consummate rupture (Lieberman, 1985). (This is the impression I gained from some conversations with his daughter, Mme Helene Veltfort, in San Francisco in 1987.)

Stekel, Rank, Ferenczi, and later Alexander are all in the tradition of psychoanalysts explicitly keen to understand the *possibilities* of creating a change in someone and *how* such *changes* operate. Some tension regarding this interest (which Freud frequently denounced by encouraging Ferenczi to concern himself with technique—e.g. in a letter on 11 January 1930) is probably due to Freud's own internal conflicts between his prime interest in understanding the human mind in himself, a model of the mind conceived within the perspective of his meta-theory, which he called metapsychology, and the need to treat people. Ferenczi (1985 [1932]) relates that Freud wrote to him, calling his patients "riff-raff" and that their main usefulness was to keep analysts alive and as learning material. This eminently relational undertaking—the psychoanalytic treatment—and the problems it poses from the point of view of *change* brought about in Ferenczi and his inheritors a theorizing based on the so-called "object relationship" and, later, on intersubjectivity. But it was not part of Freud's preoccupations and provoked tension with his theoretical constructions—a "one-body psychology" to use Balint's term. Ambiguities and rejection were the result of it.

Pluralism

The criteria for success and cure have always been intrinsically problematic; the literature has kept track of the interminable dissensions on the subject. Outside the movement, the therapeutic effectiveness of psychoanalysis has been disputed since Eysenck and Wilson (1973), and has even motivated projects as monumental as the Menninger Clinic's (Kernberg et al., 1972; Wallerstein, 1986). It continues up to Luborsky et al. (1988).

Since criteria for validating the "truth" of psychoanalytic theses have proved equally difficult and controversial (cf. the polemic around Grünbaum's, 1984, book), uncertainty about the scientific status of psychoanalysis and its methodology is like a shadow that pursues it throughout its history. Freud declared on several occasions that it was a natural science (1914d, p. 59; 1940a [1938], p. 159). At least officially, he adopted an empiricist–positivist epistemology. In 1895, he began to work on a project for *neurobiologists* (1950a [1895]) and subsequently was concerned with anthropology (1912–13), linguistics (1910e), and biology (the drives, 1905d, 1915c; the death instinct, 1920g). This began a tradition in which psychoanalysis turned sometimes to the biological sciences (e.g. with Alexander), sometimes to the social sciences, even to linguistics (with Lacan), drawing from these borderline areas, with the aim of eventually finding a scientific model. But the outcome of such a procedure is the constitution of *several scientific traditions* within psychoanalysis. Relatively "simple" models, such as Freud's way of understanding *Little Hans*, either in the *Studies on Hysteria* (1895d) or even in the "Project" (1950a [1895]), were the source of a whole development, a quasi-"behaviourist" model of psychoanalysis parallel to the present cognitive models of psychotherapy. It is characterized by an understanding of circumscribed so-called "psychodynamic" conflicts. It can be seen both in the brief psychotherapy approach (e.g. Malan) and in therapeutic interventions in children (Anna Freud). This vision of psychoanalysis demands clarity and the necessity for empirical proofs for the theses put forward (Horowitz, Luborsky, Strupp). In contrast with this, psychoanalysis is experienced as an initiation rite, a voyage to the meeting of subterranean forces intuitively grasped through dreams, an experience that is at the same time mystic and secular, capable of radically transforming the personality, and one that, at its limits, would be so inexpressible that its very narration poses practically insurmountable difficulties. The reader will be in no doubt that my position lies between these two extremes. To return to the simplicity of the first Freudian models, minimizing the whole of the subsequent development of psychoanalysis in order to reduce it to an inevitably basic and poor scientific model seems to me like renouncing the *richness* of a century's experience, which has unquestionably brought so much to the understanding of man. On the other hand, however, a radically *ascientific* position would exclude psychoanalysis from the scientific community, to its detriment and to the detriment of the experience of our knowledge of man. Furthermore, the present models, inspired partly by the development of information technology and partly by what is called artificial intelligence, enable us to apprehend semantic complexities and a wealth of processes previously unexplored. New researches on the methodological and epistemological bases of psychoanalysis after so many unfortunate and failed initiatives ought to be full of promise. For the first time in the history of this science, we have at our disposal templates that enable us to articulate our experience in the language of interdisciplinary communication and that, above all, facilitate a better understanding and formalization of our own experiences.

Freud's epistemology was to do with finding in the unconscious entities of a certain reality that needed to be confirmed by neuroanatomical and neurophysiological facts bearing a phylogenetic heritage. Since then, on the contrary, under the influence of contemporary culture, philosophy, and science, interpretation or the analysand's words and the perception of his infra-verbal communications are considered to be conditioned by childhood experience, the characteristics of the subject's development, even the biological substrates (according to Bowlby), and, more generally, by anthropological and/or linguistic conditions. In other words, the analyst's path from his initial *apprehension in the analytic situation* to his interpretation, is conceptualized by today's analysts differently from Freud and his first companions. This difference is too often hidden. Freud, certainly, looked for the Kantian "thing as such", an entity that can be grasped, although hidden, like archaeological objects. Having undergone the influences of phenomenology, hermeneutics, linguistics, and anthropology, as well as in some cases modern biology and information theory, *contemporary* psychoanalysis can only have a *different epistemology*.

Bringing this out into the open is a worthwhile scientific task, despite the often proven fears of the disruptive nature of such clarification.

Psychoanalysis has frequently oscillated between the wish to be a *transparent* discourse, public and comprehensible as it most certainly was in the writings of Freud and his pioneers (Abraham, Ferenczi), and an *esoteric* discourse for the initiated, more and more elliptical, "profound", and allusive, a jargon hindering—or avoiding—exchanges with the rest of the scientific community. (Lacan himself oscillated between wanting to be understood and wanting to participate with other influential representatives of the human sciences and being sibylline, demanding from his interlocutors an effort of comprehension and then, presumably of elaboration.) Up to now, in my opinion, this theme has been all too absent in psychoanalytic thinking, but it is a matter of a fundamental choice or attitude.

The various psychoanalytic communities (and even different groups within the same country) do not bear the same ideas and *ideological messages*. The French-speaking societies hoped to achieve a greater freedom through psychoanalysis, and this movement went together with a certain sense of superiority. In informal conversations of the initiated, the adjective "obsessional" was bandied about; it has become a pejorative term, with connotations of "pariah", "pagan", "*goy*", always rejecting the uninitiated, the outsider. Other groups felt that being considered "perverse" demarcates them in relation to the "dreadful petits-bourgeois" and would also confer some superiority. In American psychoanalysis, on the other hand, it is "being scientific" that carries the stamp of the élite. The intelligent psychoanalyst recognized by his students as a true scientist is the ideal. In Great Britain, the supreme value seemed to lie in the solid therapist. It is not surprising, therefore, that the various communities understood each other only with considerable difficulty. These collective phenomena are rarely analysed—a fact that is surprising at a time when social and political factors are constantly under close scrutiny.

There is also a noticeable *bid* for ever "*deeper*" insights. After sometimes discouraging experiences, some analysts expected from the Kleinians or Lacanians better tools than those with which they were trained. Unfortunately, there is no way of knowing the quality of the tools offered except that probably new enthusiasm gives them a power lacking in what has already been used and abused. But this is surely not a simple *renewal* of a theory.

Psychoanalytic science, like the human sciences in general, has evolved differently from what was anticipated in the 1940s and 1950s. In the present age of post-modernism, the time of grand syntheses is past. The dream of constructing a general psychology from psychoanalysis has not been realized (Loewenstein, Newman, Schur, & Solnit, 1966). Psychoanalysis has undoubtedly retained its pioneer position in the study of the expression of emotionality, especially of regressive phenomena, of fantasies and their communication in the particular dyad of psychoanalytic treatment. Equally, it has maintained some influence on developmental psychology. Other areas have been subject to rigorous research: a new biology has come into being, which, with ethology, has enriched them with new and exciting knowledge without the implementation of psychoanalysis. Experimental research, with rigorous methods of verification, has deepened our understanding of defence mechanisms, for example, just as in the information sciences it has enlarged the understanding of systems and processes, especially the processes of change. Many of these formulae have improved our capacity to test or a least to make models of Freudian insights. From this angle, psychoanalytic scientific traditions and the rich and far-reaching experiences linked to them have been able to be understood and deepened in an exchange that sometimes, discreetly, passes in silence over its origins. They are, however, often the fruit of a reciprocity between the cultural environment and psychoanalytic thinking. So can one imagine Winnicott without the influence of existentialism, or Viderman without the deepening influence of philosophical traditions? A protective attitude towards the originality of psychoanalysis has often driven authors to the discreet concealment of all sources of their thinking other than psychoanalytic ones. Jacques Lacan's work is an example of this. There are innumerable inexplicit references that do not necessarily detract in any way from its originality. Yet the various sources, especially the phenomenological and linguistic ones, deserve emphasis, since there is no doubt as to their enriching effect. The successive enrichments

of psychoanalysis are incomprehensible without this sometimes overt, sometimes covert, dialogue with ideas and knowledge springing from the cultural *environment*—since Freud.

The reconstruction of original ideas in the scientific and
therapeutic tradition instigated by Sigmund Freud and the examination of their impact today are important tasks. But there
is a divergence of opinion as to which ideas are the most original or most fundamental ones. Some models of so-called psychodynamic psychotherapies—those of David Malan (1965) or
Mardi Horowitz (1988a), for example—are based on a reconsid-
eration of the *first model* of the suppression of anxiety and the
constitution of the neurotic unconscious, and thus of *repression*. For Léon Wurmser (1987), on the other hand, the analysis
of the superego and defences in the serious neuroses—or in
other words, the exploration of guilt and censorship—remains
fundamental. Each has his own "return to Freud", his own
appreciation of what is judged to be central and worthy of exploration along traditional lines. The argument about what is
"more Freudian" seems antihistoric and at times ridiculous.
Those who consider that the analyst's function ought to be
taken into consideration in the first place sometimes deduce
that it is his *freedom* (especially to fantasize and to give free
rein to his imagination) that is fundamental. It is certainly true
that the analyst's functioning is crucial to the practice from
which the theory derives—in so far as there is a "match" to be
found in the functioning of the analysand. But this is depend-
ent on both of them meeting up in their functioning. The analyst may thus have an effect on the mental state of the
analysand. The analyst's freedom to fantasize in his self-analysis finds its limits in the process that unfolds and has as a
common frame of reference ("*a container*") the *initial pact*.

CHAPTER ELEVEN

Freud, Jung, Sabina Spielrein, and the torments of the countertransference

Biographical background

The letters that were exchanged between Sigmund Freud [1856–1939], Carl Gustav Jung [1875–1961], and Sabina Spielrein [1885–1941] are of paramount interest for the understanding of countertransference. We shall see why; but first of all let me introduce Sabina Spielrein. Born at Rostov-on-the-Don, she was the oldest daughter of a rich Jewish tradesman, Nikolay Spielrein, and his wife Eva Luyublinskaya, a trained dentist (who no longer worked after the birth of her children). Sabina became Jung's patient. She met him in 1904 at the Burghölzli, a psychiatric clinic in Zurich, where her parents had taken her for various problems. Sabina, aged 19—with "a mature figure and smooth skin", as she described herself (1980)—began an analysis with Jung, aged 30. Carl Gustav grew attached to this young, dark-complexioned, black-haired hysterical woman. Their relationship became complicated. Sabina dreamed about having a child with Jung, a "Siegfried". . . . However, despite all the complexities and what today would be called the "bad habits" of the treatment, it was an obvious *success*. In 1911 she obtained her doctorate in medicine and wrote her thesis, "On the Psychological Content of a Case of Schizophrenia", which was to appear in the *Jahrbuch für psychopathologische und psychoanalytische Forschungen* (1911). She became a member of the Zurich group of the Swiss Psychoanalytic Society as well as of the International Psycho-Analytical Association; she was the author of important articles and proposed in her subsequent works the idea, among others, of the "death instinct" (Spielrein, 1912).

At around this time, in 1909, Sabina's brother, Isaak Spielrein, was a student in philosophy and psychology at Berlin, Leipzig, and Wroclaw (he later became a professor of psycho-technology). Another brother, Jan, was in Paris and then in Stuttgart, studying mathematics and physics (he was to become an internationally renowned mathematics professor and to marry a sister-in-law of Karl Liebknecht, the famous German communist leader). The third brother, Emil, was a biologist.

In January 1912, Sabina married a Jewish Russian doctor, Pavel Scheftel. Their daughter, Renata, was born in 1913. She later became a violinist and a promising music student at the Moscow Conservatoire. Sabina embarked on studies in musical composition. After practising as a psychoanalyst in Zurich, she worked for some years in Geneva and for eight months, in 1921, she was Piaget's analyst. In 1923, four years after its foundation, her name appeared among the 36 members of the Swiss Psychoanalytic Society, and her address in Geneva was "Frau Dr. med. Sabina Spielrein-Scheftel, Pension Göbler, 6, Rue Prévost-Martin". It was in Geneva that her correspondence with Jung was found, at the Palais Wilson.

After the October Revolution, her brothers, Isaak and Jan, returned to the Soviet Union, their father's homeland, because of their ideological convictions. Sabina, in turn, after some hesitation, also returned to Rostov-on-the-Don in 1923, in the post-revolutionary years, when there was a great opening-up of the political spectrum.

She discovered a group of psychoanalysts there, and a Society, founded in 1922, whose leader was N. E. Ossipov, a psychiatrist who also trained at the Burghölzli and who, more-over, had visited Freud in 1908. He returned to Russia, where, both before and after the October Revolution, he organized the publication of a number of psychoanalytic works. Another Russian psychiatrist from Odessa, M. B. Wulff, was a member of the Vienna Society. Ms. Dr. Rosenthal, a psychoanalyst, also trained in Zurich, was responsible during the revolutionary government for the running of the Policlinic for psychoneurotic illnesses at the Institute of Cerebral Pathology directed by professor Bechterev, the well-known pathologist. A. R. Luria, the famous psychologist, took an impassioned stand on behalf of psychoanalysis at the beginning of 1920. In 1921, Vera Schmidt organized a therapeutic children's home based on August Aichorn's model in Vienna. This meant that the psychoanalytic movement in Moscow, Odessa, and Kiev seemed destined for a promising future, creating its own institutes, centres of research, and even a publishing house prepared to publish a whole series of psychoanalytic books (Luria, 1925).

Then, around 1925, the attacks on psychoanalytic concepts began, against the background of the installation of Stalinist pressure. Luria left the Psychoanalytic Society's committee in 1927, and Wulff resigned his presidency in 1928. By the time Stalin declared, in 1936, that psychoanalysis was an "idealistic" science, it was already barely existing in the Soviet Union. In a letter to Ossipov (23 February 1927) Freud prophesied that Professor Ermakov would have to moderate or firmly abandon his zeal for psychoanalysis and said clearly that even if psychoanalysis was not hostile to any party, it needed freedom to develop (Fischer & Fischer, 1977). "Pavlovianism", conditioning—the simplistic application of some of the great neurophysiologist I. P. Pavlov's ideas—came into the hands of pupils obsessed with ideology and gained the upper hand; its adherents had only ferocious contempt for psychoanalysts, some of whom had been able to escape abroad in time. One of the first, Ossipov, born in 1877, had already emigrated to Prague in 1917, and between 1923 and 1932 he became the director of the psychiatric out-patients service there and a teacher at the University of Prague. Theodor Dosuzkov, born at Baku in 1899, left Russia a year after Ossipov, in 1918, and arrived in turn in Prague in 1921, via Constantinople. He studied medicine there and became the one who was to maintain the continuity of the psychoanalytic tradition even during the Nazi occupation. He was at the origin of the expansion of psychoanalysis in Czechoslovakia after the Second World War, until the Stalinist repression in 1948. The Czech psychoanalytic society was dissolved in 1952, and its members were driven to ideological self-accusation.

Meanwhile, in the Soviet Union, things had changed. Sabina's brothers were the first in the family to suffer a tragic fate: Isaak, the professor of psychotechnology, was arrested in 1935; Jan, a famous mathematician, and Emil, a biologist, disappeared in 1937. All three were killed during the Stalinist purges.

In 1938, Pavel, Sabina's husband, died. At first Sabina survived, but Rostov was occupied by the Germans in 1941. One day she and her two daughters—Renata, by then a well-known violinist, and Eva, born in 1925—were led to the synagogue with their fellow Jews. Despite Sabina's protests in impeccable German, she was shot on June 22, along with all the other Jews of the city.

One wonders whether D. M. Thomas knew this story, which was revealed in December 1983 by McGuire after laborious research undertaken with the help of a Swedish journalist, Ljunggren, so reminiscent is it of Thomas's *The White Hotel* (1981).

* * *

The correspondence between Freud, Jung, and Sabina Spielrein gives an idea of certain aspects of the cultural life of the two Swiss cities, Zurich and Geneva, where psychoanalysis established itself at the time. We find in it, not only the Zurich of Bleuler, director of the Burghölzli, and of Jung, his assistant, but the Zurich of J. J. Honegger, a colleague and protégé of Jung at the Burghölzli, who committed suicide in March 1911, of Otto Gross, a drug addict, dreamer of a Freudian–Marxist paradise, precursor of W. Reich; of Hermann Rorschach, Karl Abraham, and others. We also find the Geneva of Claparède, Flournoy, and de Saussure.

The characters of the three protagonists are clearly distinguishable: Freud, as we know him through history; Jung, at first deferential, submissive, then rebellious; and Sabina, full of enthusiasm, a little naive, very talented, seeking and finding her identity with an impressive dignity and wisdom. Freud and others, such as Arnold Toynbee, have said that those who do not know history are condemned to repeating it. This particular history demonstrates how, little by little, the first psychoanalysts discovered that what they wanted to create—an objective medical science à la Charcot, a detached description of morbid states—*becomes a practice* in which the feelings, emotions, and affects—both the analysand's and the analyst's—play an important part. It is the discovery of the analyst's *personal involvement*, of his subjectivity, of what comes to be called the countertransference.

These three characters are revealed to us from the viewpoint of a personal correspondence, which—it should be remembered—was not destined for publication, and they only grow with closer acquaintance. One encounters them in their difficulties, sees them defending themselves, sometimes even *with blows of theory*. They become touchingly human, brothers and sisters of us all.

A deeper reflection of this sort on the history of psychoanalysis ought to teach us a great deal about its practice and about our personal involvement as psychoanalysts—for the history of a science is the science itself.

Exchange of letters

The correspondence between Freud and Jung begins like a classical drama. Among Jung's first letters one of the objects of the later drama is already revealed: "I do not think you will take the 'sexual' point of view that I am defending as excessively moderate" (an off-print, probably from *Assoziation, Traum und hysterisches Symptom* [Association, Dream and Hysterical Symptom], *Journal für Psychologie und Neurologie*, 1906) and adds: ". . . it is possible that my reservations about your far-reaching views are due to lack of experience" (Jung to Freud, 23 October 1906). The drama continues. Jung refutes the idea that he is out of line with Freud in his estimation of the role of sexuality:

> I am sincerely sorry that of all people I must be such a nuisance to you. I understand perfectly that you cannot be

anything but dissatisfied with my book [*The Psychology of Dementia Praecox*, 1907] since it treats your researches too ruthlessly. . . . The principle uppermost in my mind while writing it was: consideration for the academic German public. [Jung to Freud, 29 December 1906]

These considerations for the German establishment were subsequently reinforced and were to contribute to Jung's deplorable political position in the 1930s. Although he is "no longer plagued by doubts as to the rightness of your theory" (31 March 1907), his affirmation marks yet again the potential discord that was to become yet more dramatic. The other point of divergence will take us to the heart of the problem of countertransference: the "difficult case of a 20-year-old Russian girl student" that he mentions to Freud in one of these first letters (23 October 1906).

For the time being, however, the complicity between the two of them remains firm: Freud makes remarks about Jones, as he did in his letters to Ferenczi:

. . . he gives me a feeling of, I was almost going to say racial strangeness. He is a fanatic and doesn't eat enough. "Let me have men about me that are fat" says Caesar, etc. He almost reminds me of the lean and hungry Cassius. . . . To his mind I am already a reactionary. [Freud to Jung, 3 May 1908]

A number of events are remarked upon, Jung's treatment of Otto Gross, among others.

In 1909, the "incident" occurring in Sabina Spielrein's treatment is brought up:

. . . a woman patient, whom years ago I pulled out of a very sticky neurosis with unstinting effort, has violated my confidence and my friendship in the most mortifying way imaginable. She has kicked up a vile scandal solely because I denied myself the pleasure of giving her a child. [Jung to Freud, 7 March 1908]

He insists—falsely, as we shall see—that

I have always acted the gentleman towards her, but before the bar of my rather too sensitive conscience I nevertheless don't feel clean, and that is what hurts the most because my intentions were always honourable.

He spoke of having discovered his "polygamous components". "The relationship with my wife has gained enormously in assurance and depth." From the viewpoint of the interaction between the professional and the private man, there follows a very interesting passage concerning

> . . . a well-known American (friend of Roosevelt and Taft, proprietor of several big newspapers, etc.) as a patient. Naturally he has the same conflicts I have just overcome, so I could be of great help to him which is gratifying in more respects than one. It was like balm on my aching wound. This case has interested me so passionately in the last fortnight that I have forgotten my other duties. The high degree of assurance and composure that distinguishes you is not yet mine, generally speaking. Countless things that are commonplace for you are still brand new experiences for me, which I have to relive in myself until they tear me to pieces. [Jung to Freud, 7 March 1909]

He is discovering that understanding his own problems helps him in the analyses of others, and that the analyses can help him see more clearly within himself. The "Spielrein problem" reappears in his letter to Freud of 4 June 1909:

> In accordance with your wish I sent you a telegram this morning, framing it as clearly as I could. At the moment I didn't know what more to say. Spielrein is the person I wrote you about. She was published in abbreviated form in my Amsterdam lecture of blessed memory. She was, so to speak, my test case, for which reason I remembered her with special gratitude and affection. Since I knew from experience that she would immediately relapse if I withdrew my support, I prolonged the relationship over the years and in the end found myself morally obliged, as it were, to devote a large measure of friendship to her, until I saw that an unintended wheel had started turning, whereupon I finally broke with her. She was, of course, systematically planning my seduction, which I considered inopportune. Now she is seeking revenge. Lately, she has been spreading a rumour that I shall soon get a divorce from my wife and marry a

certain girl student, which has thrown not a few of my colleagues into a flutter. What she is now planning is unknown to me. Nothing good, I suspect, unless perhaps you
are imposed upon to act as a go-between. I need hardly say
that I have made a clean break. Like Gross, she is a case of
fight-the-father, which in the name of all that's wonderful I
was trying to cure (*gratissime*) (!) with untold tons of pa-
tience, even abusing our friendship for that purpose. . . .
Gross and Spielrein are bitter experiences. To none of
my patients have I extended so much friendship and from
none have I reaped so much sorrow. Heartiest thanks for
your blessings on my house! I take them as the best
omen.

Is this not the problem of countertransference in all its fullness? Jung rationalizes his personal involvement in the analysis: "I remembered her with special affection." "She was my
psychoanalytic training case", but, as we shall see, his sincer-
ity, his honesty is weakening. He dares not write openly to
Freud that "devoting a large measure of friendship" to Sabina
means that in reality he became *her lover* when the treatment
was finished, no doubt in the wake of feelings mobilized during
its development. Freud replies to his colleague:

Since I know you take a personal interest in the Sp[ielrein]
matter, I am informing you of developments. Of course
there is no need for you to answer this. I understood your
telegram correctly, your explanation confirmed my guess.

He cannot refrain from adopting an ambiguous attitude: after
Sabina had proposed meeting Freud in Jung's presence, Freud
writes:

Well, after receiving your wire I wrote Fräulein Sp. a letter
in which I affected ignorance, pretending to think her sug-
gestion was that of an over-zealous enthusiast. I said that
since the matter on which she wished to see me was of
interest chiefly to myself, I could not take the responsibility
of encouraging her to take such a trip and failed to see why
she should put herself out in this way. It therefore seemed
preferable that she should first acquaint me with the nature of her business. I have not yet received an answer.
[Freud to Jung, 7 June 1909]

Honesty and rupture

The whole problem of affectivity is experienced and developed in a very *dramatic* manner in this trio of the very genuine Sabina Spielrein, Jung embarrassed and constantly justifying himself, and Freud, who wants to understand. Freud's conclusion is that it is absolutely *vital* to "surmount countertransference" (Freud to Jung, 2 February 1910); not only that, but to do so "*completely*" (Nunberg & Federn, 1962; session on 9 March 1910; Binswanger also recalled Freud's statement that in every analysis countertransference must be recognized and surmounted and that only then is one able to face oneself— Binswanger, 1956; Freud's letter to B., 20 February 1913). He writes to Jung on 7 June 1909:

> Such experiences, though painful, are necessary and hard to avoid. Without them we cannot really know life and what we are dealing with. I myself have never been taken in quite so badly, but I have come very close to it a number of times and had "*a narrow escape*". I believe that only grim necessities weighing on my work, and the fact that I was ten years older than yourself when I came to psychoanalysis, have saved me from similar experiences. But no lasting harm is done. They help us to develop the thick skin we need to dominate "countertransference", which is after all a permanent problem for us; they teach us to displace our own affects to best advantage. They are a *blessing in disguise* [in italics and in English in the original].

Freud expresses here how, in what he called, in a more appropriate procedure, something endopsychic (Freud to Jung, 18 June 1909), this experience ought to be able to be transformed into another *constellation* through an *intrapsychic* modification of feelings mobilized in the transference, and the enrichment of the person through the mechanisms of introjection and working through.

Such a conclusion is not, however, easily reached. It is noticeable that he twice uses expressions in English in the letter quoted, just as earlier, in his first description of Oedipal feelings to Fliess (3 October 1897), he had recourse to the Latin "*matrem nudam*". It reflects his difficulty in dealing with these feelings and affects. He is on the brink of hypocrisy when he writes to Jung (18 June 1909):

> Spielrein has admitted in her second letter that her business has to do with you; apart from that, she has not disclosed her intentions. My reply was ever so wise and penetrating; I made it appear as though the most tenuous of clues had enabled me, Sherlock Holmes-like, to guess the situation (which of course was none too difficult after your communications) and suggested a more appropriate procedure, something endopsychic, as it were. Whether it will be effective, I don't know. But now I must entreat you, don't go too far in the direction of contrition and reaction. Remember Lassalle's fine sentence about the chemist whose test-tube had cracked: "With a slight frown over the resistance of matter, he gets on with his work." In view of the kind of matter we work with, it will never be possible to avoid little laboratory explosions. Maybe we didn't slant the test-tube enough, or we heated it too quickly. In this way we learn what part of the danger lies in the matter and what part in our way of handling it.

Freud is resorting to science as a means of protection. He seems to be seeking a balance between affects that have been mobilized and the scientific mastery that enables him to keep the analytic process on an "endopsychic level": a love affair, yes, but one that ought to be able to be transformed into a matter of understanding and introjection. (Following this, Jung is very grateful, and their relationship is temporarily strengthened—Jung to Freud, 12 June 1909.)

It is impossible to read Sabina's letters without emotion. Emma Jung and Sabina Spielrein (like Gizella, Ferenczi's wife) are women of astonishing sensitivity and very genuine, too. Sabina hopes that Jung will show himself worthy of love (Spielrein to Freud, 10 June 1909). She writes:

> Four and a half years ago Dr. Jung was my doctor, then he became my friend and finally my "poet", i.e., my *beloved*. Eventually he came to me and things went as they usually do with "poetry" . He preached polygamy; his wife was supposed to have no objection, etc., etc. [italics added]

An incredible story follows, in which Sabina's mother receives an anonymous letter (letter from Spielrein to Freud, 11 June 1909):

> There is reason to suspect his wife. To make a long story short, my mother writes him a moving letter, saying he had saved her daughter and should not undo her now, and begging him not to exceed bounds of friendship. Thereupon his reply: "I moved from being her doctor to being her friend when I ceased to push my own feelings into the background." His argument is as follows: "I could drop my role as doctor the more easily because I did not feel professionally obligated, for I never charged a fee." He concludes: "My fee is 10 Frs. per consultation. . . . I advise you to choose the prosaic solution."

(The recourse to money is compensation for the frustration of the analyst.)

Then comes Jung's letter to Sabina's mother (12 June 1909):

> When this occurred, I happened to be in a very gentle and compassionate mood, and I wanted to give your daughter convincing proof of my trust, my friendship, in order to liberate her inwardly. That turned out to be a grave mistake, which I greatly regret.

The disastrous effect of this affair on Sabina is obvious and one can only admire her nobility of soul. She writes again on 12 June 1909 that her love is transcendent, and a year later, in 1910: "And now? He is close to me again. At least (if I am so fond of him) I could give him a little boy, as we used to dream of? Then he could go back to his wife. Yes, if only it were that easy" (Carotenuto, 1980, p. 13).

There is a postscript to this story from Jung's pen (Jung to Freud, 21 June 1909):

> I have good news to report of my Spielrein affair. I took too black a view of things. After breaking with her I was almost certain of her revenge and was deeply disappointed only by the banality of the form it took. The day before yesterday she turned up at my house and had a *very decent talk* with me, during which it transpired that the rumour buzzing

about me does not emanate from her at all. My ideas of reference, understandable enough in the circumstances, attributed the rumour to her. . . . She has freed herself from the transference in the best and nicest way and has suffered no relapse (apart from a paroxysm of weeping after the separation). . . . Although not succumbing to helpless remorse, I nevertheless deplore the sins I have committed, for I am largely to blame for the high-flying hopes of my former patient. . . . I imputed all the other wishes and hopes entirely to my patient without seeing the same thing in myself. When the situation had become so tense that the continued perseveration of the relationship could be rounded out only by sexual acts, I defended myself in a manner that cannot be justified morally. Caught in my delusion that I was the victim of the sexual wiles of my patient, I wrote to her mother that I was not the gratifier of her daughter's sexual desires, but merely her doctor, and that she should free me from her. In view of the fact that the patient had shortly before been my friend and enjoyed my full confidence, my action was a piece of knavery which I very reluctantly confess to you as my father.

On the 12 October 1911, Freud recounts that Fräulein Spielrein "turned up unexpectedly" at his home. A page is turned; it is the start of Freud's relationship with Spielrein and the beginning of the end of Jung's. The scene changes: it is Freud who, referring to a meeting of the Viennese Society, considers now that "Fräulein Spielrein . . . was very intelligent and methodical" (Freud to Jung, 12 November 1911) and a few days later ". . . she is rather nice, and I am beginning to understand" (Freud to Jung, 30 November 1911).

At this point—and perhaps also because of that episode—Jung's attitude towards Freud changes. Is it a reaction to this evidence of his weakness? There is little doubt about it. On 18 December 1912 he writes: "I admit the ambivalence of my feelings towards you, but am inclined to take an honest and absolutely straightforward view of the situation." He moves in to attack:

If you doubt my word, so much the worse for you. I would, however, point out that your technique of treating your pupils like patients is a *blunder*. In that way you produce either slavish sons or impudent puppies (Adler, Stekel and the whole insolent gang now throwing their weight about in Vienna). I am objective enough to see through your little trick. You go around sniffing out all the symptomatic actions in your vicinity, thus reducing everyone to the level of sons and daughters who blushingly admit the existence of their faults. Meanwhile you remain on top as the father, sitting pretty. For sheer obsequiousness nobody dares to pluck the prophet by the beard and inquire for once what you would say to a patient with a tendency to analyse the analyst instead of himself. You would certainly ask him: *Who*'s got the neurosis? You see, my dear Professor, so long as you hand out this stuff I don't give a damn for my symptomatic actions; they shrink to nothing in comparison with the formidable beam in my brother Freud's eye. I am not in the least neurotic—touch wood! I have submitted *lege artis et tout humblement* to analysis and am much the better for it. You know, of course, how far a patient gets with self-analysis: *not* out of his neurosis—just like you.

Voices are raised, and a dispute begins as to whose—Jung's or Freud's—neurosis it is (Jung to Freud, 3 December 1912):

My very best thanks for one passage in your letter, where you speak of a "bit of neurosis" you haven't got rid of. This "bit" should, in my opinion, be taken very seriously indeed.

He makes the same reproach that is repeated in his autobiography (1962): "Our analysis, you may remember came to a stop with your remark that 'you could not submit to analysis *without losing your authority*'". The tension continues to rise. Freud writes a serious reply on 5 December 1912:

You mustn't fear that I take your "new style" amiss. I hold that in relations between analysts as in analysis itself every form of frankness is permissible. I too have been disturbed for some time by the abuse of psychoanalysis to which you refer, that is, in polemics, especially against new ideas. I do not know if there is any way of preventing this entirely; for the present I can only suggest a household remedy: let each of us pay more attention to his own than to his neighbour's neuroses.

But the break is inevitable. On 6 January 1913 Jung declares:

I accede to your wish that we abandon our personal relations, for I never thrust my friendship on anyone. You your-
self are the best judge of what this moment means to you.
The rest is silence.

On 1 January 1913 Freud writes to Sabina Spielrein that his
relationship with her "Germanic hero" is definitely severed. He
describes Jung's behaviour as too dreadful and relates how his
opinion of Jung was quite changed after Sabina's first letter (to
Freud). However, as far one can see, Jung's collaboration on
the professional level continues. In August of the same year
Freud writes to Sabina Spielrein again that he will not pass on
her greetings to Jung, as she well knows. The issue of the
Jahrbuch (Vol. 5, No 2) that came out at the end of 1913 has the
following on p. 757 (with the signature of C. G. Jung):

> I have found myself obliged to resign as editor of the
> *Jahrbuch.* The reasons for my resignation are of a personal
> nature, on which account I disdain to discuss them in pub-
> lic.

And in April 1914:

> The latest developments have convinced me that my views
> are in such sharp contrast to the views of the majority of
> the members of our Association that I can no longer con-
> sider myself a suitable personality to be president. I there-
> fore tender my resignation to the council of the presidents
> of the branch societies, with many thanks for the confi-
> dence I have enjoyed hitherto.

The reasons for the break are obviously complex. Jung's
countertransference to Sabina Spielrein and his defeat vis-à-
vis Freud played an important part. Jung, with his injured self-
esteem and guilt, did not tolerate Freud's viewpoint. But he
challenged Freud in areas that open up problems marking the
history of psychoanalysis—particularly the question of relationships between analysts.

The study of this exchange reveals the difficulties raised by
the countertransference at one of the historic moments of its
discovery and understanding, and the complexities of the relationship between the two men. Some of Jung's contributions
have been ignored—his views clarifying psychosis, for example,

rediscovered in part much later by Melanie Klein. Afterwards, he accomplished his own work outside the Freudian circle—a work of cultural interpretation inspired by a particular phenomenology. The problem of the psychoanalytic impact on the person of the analyst and of the dialogue between analyst and analysand were taken up both then and later by Ferenczi and his followers.

CHAPTER TWELVE

Freud
and his favourite disciple

I am fairly generally regarded as a restless
spirit, or, as someone recently said to me at
Oxford, the *enfant terrible* of psychoanalysis.

Sándor Ferenczi, *Final Contributions*, p. 127

F reud's concern was to construct what today can be
called an *anthropological* theory. He tried to clarify the
functioning of the human mind through models drawn
from the *natural sciences* (e.g. Freud, 1914d, p. 59; 1940a
[1938], p. 159). His conceptualization, as we know, is based on
the hypothesis of forces hidden behind the various manifestations of the mind. He believed the most important of these to be
the *drives* and the forces that are in conflict with them (censorship, repression, superego, etc.). In the framework of his notion

A different text on the same subject has previously been published
in collaboration with Ernst Falzeder in German [*Jahrbuch der Psychoanalyse*, 24 (1989): 109–127], and in English [*Free Associations*, 2/1
(1991): 1–20].

of a dynamic unconscious, he endeavoured to identify precisely those forces that operate without the subject's knowledge. Around 1900–1905, his interest became centred on the *drive*. Gradually, the notions of transference and countertransference gained in importance and began to facilitate a grasp of how these forces operate, *how* they can be understood in the interaction, in the *present*, and of their hidden *action* in the individual's past *history*. As a result of this change of emphasis, the prevailing model progressively became one of *interactions* (including affective interactions). Ferenczi is the father of those psychoanalysts whose *principal* interest is with this interaction, but *not* to the exclusion of the underlying drives. (On the other hand, others in the history of psychoanalysis more or less explicitly renounced the drive model. At present, many analysts think that the concept of drive has been, or should be abandoned, others fear that this might sever the links between psychoanalytic theory and biology.) The interactional model is favoured by the majority of analysts who are successors to Ferenczi and the Budapest school, particularly the English school inspired by Melanie Klein, who was also a pupil of Ferenczi (curiously, only her affiliation with Abraham is customarily mentioned, omitting the one with Ferenczi—with certain exceptions such as Grosskurth, 1986), the Middle Group (Michael Balint, Donald W. Winnicott, Wilfred Bion, and, prior to the later influence of Michael Balint, Ella Freeman Sharpe and Margaret Little), and in another historical line (Harry Stack Sullivan, Clara Thompson, Harold Searles, Robert Langs). Interaction inevitably implies *language*. The language and communication of the analysand is one of the most important elements in the psychoanalytic interaction.

The role attributed to linguistic manifestations takes us back to the difference between Lacan's conception, centred exclusively on the verbal, and the position taken by the majority of the other schools, where interaction is understood in a wider sense, encompassing the non-verbal as well as the verbal, text and context, the semiotic and also the pragmatic, to be taken into consideration in what can be called "psychoanalytic *communication*".

What is occurring is a phenomenon that parallels contemporary sciences in general—namely, a move from the dominant paradigm of *energy* to one of *information* (the movement and transformation of symbols). In psychoanalytic theory this corresponds to the move from the concept of drive energy to that of information exchange (cognitive and emotional) in the *relationship* (verbal and non-verbal communication).

At the heart of the problem lies the conviction that what is *experienced* in the analysis can contribute to changing the course of the "repetition compulsion". This brings in the question of how much should be accorded to non-specific factors of *change*—factors such as the setting, the analyst's personality, as opposed to specific interpretations of the Oedipus complex, of infantile sexuality, etc.

Rethinking the epistemological foundations of psychoanalysis—especially those of its theory—in the light of contemporary human sciences, undoubtedly has the advantage of being able, at the end of this century, to reinstate psychoanalysis as a scientific discourse. Such a step should result in presenting to the scientific community a credible account of the contribution of psychoanalysis to the understanding of man and his mental functioning. On the other hand, does such a re-evaluation in the light of the contributions of contemporary human sciences represent an intolerable revision—even a betrayal—of psychoanalytic theory? This is a question raised by those brought up in the scientific tradition of the psychoanalytic community. In any case, since Freud, psychoanalysis has always sought dialogue with the related biological and human sciences (and, with the development of ideas, this has often resulted in its accepting innovations, often in a less radical, more attenuated form than the original—historical examples are those of Rank on the importance of separation, and the majority of Melanie Klein's ideas). The dialogue psychoanalysis maintained with contemporary linguistics, ethnology, and biology is evidence of this. Disregard of this fact would underestimate the influence of contemporary linguistics and phenomenological and existentialist philosophies on the psychoanalytic theory of our time—for example, in French-speaking countries the work of Lacan and others of his school. Obscure and inexplicit references are never fruitful in the development of science. Clarifying them and seeking to renew the dialogue with other allied fields seems to me important in highlighting the fundamental contributions

of psychoanalysis—both its clinical experiences and its theory—
in the search for an understanding of man, particularly of contemporary man. This is also important to its survival, to the
possibility of ensuring for it a role in the development of human
thinking, and for preserving for future generations the richness
of its methodology and of the ideas that govern it.

A dialogue beyond death

It is true that whenever a crisis broke out, Freud invariably
showed himself what he really was, a truly great man, who
was always accessible and tolerant to new ideas, who was
always willing to stop, think anew, even if it meant re-ex-
amining even his most basic concepts, in order to find a
possibility for understanding what might be valuable in
any new idea. It has never been asked whether something
in Freud has or has not contributed to a critical increase of
tension during the period preceding a crisis. Still less has
any analyst bothered to find out what happened in the
minds of those who came into conflict with Freud and what
in their relationship to him and to psychoanalysis led to the
exacerbation. We have been content to describe them as
the villains of the piece. . . . Maybe Rank's case is less suit-
able for this examination but I am quite certain in
Ferenczi's case one could follow the development which,
prompted by the characters of the two protagonists, led to
the tragic conflict. [letter from Michael Balint to Ernest
Jones, 31 May 1957—Balint Archives, Geneva]

Psychoanalysis is still, to this day, affected by the difficulties in the relationship between Freud and Ferenczi. How is this
to be explained? On the one hand they themselves did not
readily agree on important theoretical and practical questions—e.g. about the nature of trauma, the relationship between internal and external reality, the granting or frustration
of satisfactions in therapy, transference and countertransference, and the nature of infantile sexuality. On the other
hand, the unwillingness to have an unbiased investigation into
the nature of this relationship and its concomitant difficulties

has disturbed the subsequent development of theory and practice and cleared the way for "over-eager salesmen" (for the history of this expression, see Haynal, 1987a) and those eager to
establish methods independently "with their own cannon". This
last expression refers to one of Freud's favourite stories (1950c,
p. 56):

> Itzig had been declared fit for service in the artillery. He was
> clearly an intelligent lad, but intractable and without any
> interest in the service. One of his superior officers, who was
> friendly disposed to him, took him on one side and said to
> him: "Itzig, you're no use to us. I'll give you a piece of ad-
> vice: buy yourself a cannon and make yourself independ-
> ent!"

Freud was attracted by uncompromising and intelligent people,
especially if they were interested in serving in his "wild army"
(Freud, 1960a, letter to Groddeck, 5 June 1917, p. 36).

All this contributed to the split between Freud and Ferenczi,
forcing their followers to identify with the one and declaring the
other wrong, dangerous, or even mad. What distortion of reality
must there have been to operate such a split, when the two
protagonists themselves never took such clearly delineated
positions! They touched on important fundamental issues—
important even for us today—in the theory and practice of psychoanalysis. Those questions created deep-rooted, even tragic
conflicts between them. Yet, it is to our advantage not to try
and reconcile too hastily the positions taken by Freud and
Ferenczi; such haste would result in a failure to understand the
fundamental nature of this controversy.

Between Freud and Ferenczi there existed a *dialogue*, a
friendship, even "an intimate community of life, of feelings, and
of interests" ["*Innige Lebens-, Gefühls- und Interessengemein-
schaft*"] (Fr. to Fer., 11 January 1933). In the scientific field,
they constantly communicated their thoughts and plans to
each other. Their influence on each other continued far beyond
their estrangement and death: Ferenczi's *Clinical Diary* (1985
[1932]), which can be looked upon as a product of "scientific
immersion in a kind of 'Poetry and Truth' [Goethe's memoirs]"
["*Versinken in eine Art wissenschaftliche 'Dichtung und
Wahrheit'*"] (Fer. to Fr., 1 May 1932), can be read as a letter

addressed to Freud. A quarter of a century after the fragments of analysis that Ferenczi had made with Freud, the latter was still preoccupied by the question of whether he himself had behaved correctly (Freud, 1937c). One of Freud's very last notes on "The Splitting of the Ego in the Process of Defence" (1940e [1938]), in which he writes that he does not know whether what he wants to convey should be considered as something that has been known for a considerable time and is self-evident, or as being completely new and apparently strange, was a central theme in Ferenczi's work during the last years of his life (Ferenczi, 1932 [308, in 309], 1985 [1932], etc.).

Freud's influence on Ferenczi's work is evident. The correspondence and the *Clinical Diary* both clearly show the extent to which he always wrote for Freud (see Fer. to Fr., 23 May 1919, 15 May 1922, 30 January 1924), how he discussed the content, timing, and placing of a publication, and how much—as we will see later—his experiences and technique were bound up in this relationship. What is less well known is the fact that the ideas expressed in Ferenczi's main work, his theory of genitality ("Thalassa", 1924 [268]), had been the subject of intense discussions between him and Freud long before its publication, and there was even a plan to write a joint work on Lamarck (Fer. to Fr., 2 January 1917).

Freud made use of other people's thinking in different ways. Although, as we have seen, he adopted many ideas from other writers, he worked through and "digested" them until they resurfaced as his own. "I have a strong tendency to plagiarism" he wrote to Ferenczi (Fr. to Fer., 8 February 1910). Thus many of Ferenczi's ideas and concepts reappeared in Freud's work, often after a prolonged latency period, and intermingled with his own ideas—for example, thoughts on homosexuality, paranoia, phylogenesis, trauma, transference and countertransference, development of the ego, technique, and parapsychology.

In addition to the scientific ties, there were other, more complex and profound ones between the two men. There was the hope fostered by Freud of a marriage between his daughter Mathilde and Ferenczi; there was the voyage to America undertaken with Jung; there were numerous common holidays, with all their concomitant pleasures and difficulties. There was Ferenczi's "attempt at an analysis" with Freud, and Ferenczi's relationship with his future wife, Gizella, and with her daughter, Elma—a relationship in which Freud had been involved, as we shall see, in various ways, amongst which was a "piece" of analysis with Elma (Haynal, 1987a). In addition, their relationships with other analysts—with Jung, Rank, Jones, Groddeck, Abraham, Eitingon, Reich—also had a part to play in the conflictual history of psychoanalysis.

Freud and Ferenczi visited one another. Ferenczi acted as host to Anna Freud in Hungary, he procured foodstuffs for the Freud family, and he referred patients to Freud during and after the war. Freud kept Ferenczi closely in touch with news of his family and of his sons during the war. They wrote about all manner of personal things—about cigars and flour, about the restriction on morning showers, on the political situation, about mutual friends and acquaintances. They wrote about Ferenczi's military service and the lack of fuel in post-war winters, about financial problems and Freud's grandchildren. Months before their travels, they began to study Baedeker and train and boat timetables.

At the same time the relationship was one of conflict, one in which offence was taken and misunderstandings led to veritable *controversies*. In 1910, for example, during their holiday together in Palermo, Ferenczi refused to allow Freud to dictate his notes on the Schreber case. For the rest of the journey the two men were unable to discuss the incident and its affective significance. Freud criticized Ferenczi, both openly or by allusion, for his relationship with Gizella and Elma Palos. It annoyed him that Ferenczi "compromised himself so deeply" with Rank (Fr. to Fer., 12 October 1924), fearing that his "paladin and secret grand vizier" was "taking a step towards the creation of a new oppositional analysis" (Fr. to Fer., 13 December 1929). He reproached Ferenczi for going "in all directions" that to him, "seemed to have no desirable aim" (Fr. to Fer., 18 September 1931), and could imagine Ferenczi's reaching maturity only after he had found "a way back" (ibid.). Freud's ironic and mordant criticism of Ferenczi's "technique of the kiss" (Fr. to Fer., 13 December 1931) is well known, and he considered Ferenczi's creative regression as a game with "dream children" on an "imaginary isle" from which he could only be snatched by "radical treatment" (Fr. to Fer., 12 May 1932). In the end, he did not hide his bitterness:

> In the last sentence of your letter [Ferenczi had written: "Greater courage and a more open expression on my part would have been an advantage"], you really accuse yourself, and I can only agree with you. For two years, you have systematically turned away from me, having probably developed a personal hostility which goes much deeper than it might appear. [Fr. to Fer., 2 October 1932]

For his part, Ferenczi reproached Freud for not having, in his analysis, "penetrated the negative feelings and fantasies, allowing them to be abreacted instead" (Fer. to Fr., 17 January 1930), and he disapproved of Freud's neglect of the process of cure (ibid.). He objected to Freud's diagnosing as an "illness" what was in fact his profound commitment to therapeutic problems (Fer. to Fr., 19 May 1932) and described the "depth of my shock" (Fer. to Fr., 27 September 1932). At their last meeting, conflicts arose between them about the paper Ferenczi had proposed to give at the approaching Wiesbaden Congress on "The Confusion of Tongues between Adults and the Child" (Ferenczi, 1933 [294]). At the end of that visit, Freud did not even shake hands with him by way of saying good-bye. Shortly afterwards, Ferenczi, already marked by the illness from which he was to die, wrote a note "On Shock": "Perhaps even the organs which secure self-preservation give up their function or reduce it to a minimum" (1932 [308 in 309]). Ferenczi seems to criticize Freud rather more in his *Clinical Diary* than in the correspondence (Ferenczi, 1985 [1932]; see also J. Dupont's introduction). But it is also a fact that, when the unpublished papers of Ferenczi were presented to him after Ferenczi's death, Freud expressed "his admiration . . . for the ideas which hitherto had been unknown to him" (Balint, 1969, in: Ferenczi, 1985 [1932], p. 14).

Is it not pretentious to reduce such a relationship to a simple "transference relationship", or to attribute to either one of the protagonists the "blame" for the derailments in the dialogue? Whatever his divergences of opinion with Freud, to label the founder of the IPA a "dissident" and think the problem is thereby resolved seems to me an error of judgement. Their controversy did not end in desertion and enmity: "the disputes between us . . . can wait. . . . I am much more concerned that you recuperate your health", wrote Freud (2 April 1933), a few weeks before Ferenczi's death. But it could no longer be resolved.

The friends of his friends

The relationship between Freud and Ferenczi was one of friendship and discord—and psychoanalysis was always interwoven in it in theory, technique, or movement, but also as personal experience.

Freud, Ferenczi, and Jung analysed each other during the voyage to America. We know what a deep impression was made on Jung by Freud's refusal to allow himself to be analysed beyond a certain point (Jung, 1962). (For a discussion on the relationship between Jung, Spielrein, and Freud see chapter eleven.) Freud analysed many of his pupils—for example, Eitingon (see Jones, 1955, p. 34)—often briefly, during walks. He even analysed his own daughter Anna ("Annerl's analysis is going very well, that apart the cases are uninteresting", he wrote to Ferenczi on 20 October 1918).

On the other hand, he refused to analyse some people, for instance Tausk (Roazen, 1969), Federn (Roazen, 1971/1976), Reich (ibid.), and Otto Gross, against whose treatment "my egoism or perhaps rather my legitimate defence has rebelled", as he wrote to Jung (21 June 1908). Groddeck (Groddeck, 1974) and Ferenczi (Fer. to Fr., 26 February 1926, 1 March 1926) each offered to take Freud into analysis, but he refused. Ferenczi and Groddeck analysed each other (see Groddeck, 1974, p. 82; Ferenczi & Groddeck, 1982). Ferenczi analysed Jones (correspondence, passim), Freud analysed the latter's friend Loe (ibid.), and they corresponded about this. Ferenczi thought that Freud ought to analyse Jung (Fer. to Fr., 20 January 1912) and vehemently disagreed with Otto Rank, who, in connection with the controversies aroused by his "birth trauma" (1924), had expressed the opinion that it was an advantage not to have been analysed (Fer. to Fr., 1 September 1924).

On 14 July 1911 Ferenczi wrote to Freud that he had taken into analysis Elma, the daughter of his friend Gizella (then married to Géza Palos), whom he had already analysed (Fer. to Fr., 30 October 1909). In the course of the analyses Ferenczi falls in love with Elma Palos, who "enters victorious into my heart" (Fer. to Fr., 3 December 1911). It is with insistence that he begs Freud to take Elma into analysis. Although reticently, Freud accepts. As it progresses, he reports to Ferenczi the details of the treatment, especially with regard to the question of knowing whether Elma's love for Ferenczi "stands up" to the analysis. Everyone concerned commits indiscretions: Ferenczi sends Freud copies of letters from Elma, "in which she *absolutely* must know what you (Freud) have written to me about her" (Fer. to Fr., 18 January 1912); Freud writes confidentially to Gizella about Ferenczi (Fr. to Fer., 17 December 1911)—and, naturally, Gizella shows the letter to Ferenczi. Furthermore, Elma's father, whom she had kept up to date with the details, wants to interfere. Ferenczi visits Freud in Vienna to talk about Elma. This meeting is kept secret from Elma, who lives in Vienna. Although she wants to continue the analysis, Freud breaks it off, for she would have reached "the narcissistic level" (Fr. to Fer., 13 March 1912). Back in Budapest, Elma is again in analysis with Ferenczi. In that way he hopes to reach some certainty about her feelings for him, and he resists her "tokens of affection" (Fer. to Fr., 27 May 1912). He remains abstinent "in a rather cruel way" (Fer. to Fr., 18 July 1912) and yet does not manage to reach a clear understanding of their respective feelings. On the other hand, "poor Elma has really no pleasure [in this analysis]" (Fer. to Fr., 26 July 1912). In the end, Ferenczi abandons Elma's analysis. [In "Psycho-Analysis and Telepathy" (1941d [1921]), Freud describes a very similar triangular relationship, in which there is also a girl who is "driven" to an analysis because a man could not decide between her and the mother.] She marries an American named Laurvik—a marriage that does not last for very long. Several years after these events, in 1919, Gizella and Sándor Ferenczi married. On their way to the registry office they learned that

Gizella's ex-husband, Géza Palos, had died (Haynal, 1987a, pp. 63 et seq.).

For years afterwards, Ferenczi suffered the sequelae of these events; he complained of depressions and hypochondriacal symptoms and had great difficulty in regaining his equilibrium. All this is characteristic of his temperament and testimony to his unrestrained involvement, without any "insurance", in the therapeutic situation and with few clear or defensive boundaries between his professional and private life. Not only was Ferenczi the analyst, but the whole man was engaged in these relationships, going to his limits with a courage that was both his strength and—in this instance—undoubtedly also his weakness.

We see Freud torn between a deep sympathy for Ferenczi's destiny and for his and Gizella's fate—a sympathy that drove him to intervene—and his doubts as to the outcome of such an intervention. He was "concerned at linking the fate of our friendship with something else indefinable" (Fr. to Fer., 21 April 1912) and wrote about "the danger of personal alienation caused by analysis" (Fer. to Fr., 4 May 1913). He saw the dangers more clearly than did Ferenczi, but despite this he analysed Elma and later Ferenczi. On one occasion he described himself as a "sentimental ass whom even old age did not prevent from making a fool of himself" (Fr. to Fer., 23 January 1912). Another time he described himself as "hardhearted", albeit "out of compassion and softness" (Freud to Gizella, 17 December 1911).

In other relationships with followers and patients alike, Freud hesitated between two attitudes—the one "sentimental" and the other "hard", between one form of relationship, which he called unexpressed transference ("repetition", 1914g), in which he allowed considerable closeness—"because I enjoy them very much" (Fr. to Fer., 6 June 1910)—gave presents and issued dinner invitations (Haynal, 1987a, p. 19) and did little if any analysis, and another form of relationship, in which he "tended towards intolerance" with regard to "neurotics" (Fr. to Fer., 20 January 1930), could remain distanced, "thwart their intrigues" (Fr. to Fer., 20 July 1912), and could "cool off" appreciably when transference wishes arose (Fr. to Fer., 24

March 1912). His position was a difficult one; many were disappointed at not finding in him the image of the "psychoanalytic superman" that they had "constructed" (Fr. to Fer., 6 October 1910), and whatever he did, he always encountered critics.

For Freud, as for Vischer, whom he readily quoted: "morality is self-evident" (Freud, 1905a). With those people for whom this was not the case, he neither could nor would adopt a psychoanalytic attitude, but he passed judgement on them—and who should deny him the right?—according to a morality that was indisputable in the eyes of the man, the doctor, the professor, the father of a family, and the founder and leader of a new movement in the Vienna of his time. Thus, for example, Stekel was, in his eyes, "an offence against all good taste" (letter to Jung, 27 April 1911), and there were things that "a gentleman should not do, even unconsciously!" (a remark Freud made about Jung, when the latter forgot to communicate to Jones in good time the date for a meeting—according to Jones, 1959, p. 154).

Freud certainly put his followers and patients in something of a dilemma when on the one hand he did not allow certain aspects of his role to be discussed and, on the other, emphasized that the partner should "tear himself out of the infantile role" (Fr. to Fer., 2 October 1910). Both Elma and Ferenczi found themselves in this same dilemma. Ferenczi could not accept that one part of a relationship may be expressed and another part not, because it was "self-evident". According to him, everything could and should be spoken about between analytically trained people. "Imagine what it would mean, *if one could speak the truth to everyone*, to one's father, to one's teacher, to one's neighbour and even to the king" (Fer. to Fr., 5 February 1910). Psychoanalysis seemed to him a means of coming into the open and expressing even the hidden and inexpressible.

Perplexed and distraught by the complications linked to his involvement, Sándor Ferenczi resorted again and again to analysis, hoping thereby to find an objective method for elucidating human relationships. And so he came to believe that with no matter what patient, every fully trained analyst ought to reach the same unmistakable and objective conclusions and would adopt the same tactical and technical measures (1928 [283], p. 89). He, too, compared the process during analysis with the chemical reactions in a test-tube (Fer. to Fr., 21 April 1909). But no analysis can manage to isolate that chemically pure part of feeling "uninfected" by the transference or the neurosis—neither in Elma's analysis with Ferenczi, nor in Ferenczi's with Freud. On the contrary, instead of the relationships being simplified, they became more complicated.

It was precisely in these episodes in his dual roles of analyst *and* analysand that Ferenczi experienced very painfully that psychoanalysis is not an instrument independent of the person who uses it. These events probably contributed to his increasing understanding that the analyst's attitude was a *variable* in the therapeutic equation, and for that reason he put it at the centre of his interests. How distressing it must have been for him, in fact, in that network of relationships between Freud, Elma, and himself, not to be able to distinguish "transference" from "real" feelings. How he must have suffered in distributing the roles of analyst, analysand, lover, friend, and follower, and how he threw himself wholly into those relationships! It is indeed understandable that he was able to see with great acuity how patients might sometimes suffer from the "hypocrisy" of the analyst's total abstinence (Ferenczi, 1933 [294], p. 127).

There is, undoubtedly, a logical path that leads from his personal to his technical experiences, to active therapy, and to the method of relaxation, and so to mutual analysis and the theoretical conceptions in "The Confusion of Tongue between Adults and the Child" (Ferenczi, 1933 [294]), and hence to the role of adults, the psychic atmosphere during the child's development, and the sequelae of trauma.

Professional life—private life

It is tempting nowadays to appraise the intricacies of the relationship between Ferenczi and Freud, their use and abuse of psychoanalysis, their indiscretions and acting out, from what could be considered our "superior" standpoint. As if today's analysts, with all their training, their training analyses and supervision, and with all the theoretical and technical equipment at their disposal, were better able to separate or optimally fuse their *professional identities* and their *private lives*! Yet, how many analysts still sleep with their patients (Fischer, 1977) and how many marry them, and how much between parent and child generations of analysts can neither be verbalized nor worked through!

Analysis is, in Freud's words, "at base, *a cure through love*" (letter to Jung, 6 December 1906—italics added). (There is a similar idea—"our treatments are treatments of love"—in another context: namely, in the Minutes of the Viennese Psychoanalytic Society, in which Freud explains the degree to which the "power" of treatment resides in love—that is, transference love: "The patient is obliged to give up his resistances through his love for us"—Nunberg & Federn, 1962, Discussion at the Scientific Session, 30 January 1907.)

For Freud, the term "love" means "transference love"; for Ferenczi it means "countertransference love". (One of Ferenczi's late followers, Leopold Szondi, from Zurich, often spoke of the therapist as a "soul donor"—an analogy with "blood donor".) Freud maintained that the attempt to work with pale copies of emotions fails, "for when all is said and done it is impossible to destroy anyone *in absentia or in effigie*" (Freud, 1912b, p. 108), unless one is behaving like the less-than-potent man who said to his wife after the first intercourse on their wedding night: "There, now you know about it; the rest is always the same" (Fr. to Fer., 20 January 1930).

Why do we tend to think that technique was less developed in Freud's day than it is today? It is as if we now had at our disposal a definite technique that is above all suspicion, as if only the first generation of analysts ever found themselves in what Ferenczi called a stage "of experimentation". In fact, one might as well say that the analysts of that time were *conscious* that they were *always experimenting*, and that from the moment we began to speak of a classical technique, we entered a stage of *illusion*, an illusion that there exists a technique that one needs only learn and apply "correctly" and on which "textbooks" can be written.

To the limits

And so to 1932, the year before Ferenczi's sixtieth birthday. He was just 35 when he made Freud's acquaintance. Behind them lay 24 years of intense friendship, decades of torments and joys in the life of a psychoanalyst who tried to go as far as possible with his method, who tried to understand himself and to understand his analysands with a passion that some, Freud included, believed necessary to denounce as exaggerated, because it was the wish to help *to the limits* of what is possible.

Ferenczi took the decisive step: he no longer tested his practice in the mirror of his master's approval or disapproval, and he decided to undertake, in the form of a diary, as exhaustive an exploration as possible. Although this diary, which covered the nine months from 7 January to 2 October 1932, is a step towards self-assertion and represents primarily a will to understand the analyst's position more thoroughly, without, this time, the to-and-fro of a correspondence, nevertheless the transference figure of Freud, to whom the diary is implicitly addressed, is clearly discernible. Furthermore, we know, through Michael Balint, that after Ferenczi's death Freud read it with great interest and respect (Balint, 1969, in: Ferenczi, 1985 [1932], p. 14).

Such a profound self-exploration should not be weighed or judged according to criteria of orthodoxy or heterodoxy—whatever is meant by these words—but on the really fundamental questioning of an analyst's life. He puts the problem at the outset: the analyst's *insensitivity*. It comes up immediately: the "real countertransference" of the analyst, the need to know more about it, and the almost caricatured idea of *mutual analysis*. Ferenczi (1985 [1932]) retraces the connection between the two:

> Mutual analysis will also be less terribly demanding, will promote a more genial and helpful approach in the patient, instead of the unremittingly all-too-good, selfless demeanour, behind which exhaustion, unpleasure, even murderous intentions are hidden. [p. 16]

The *body* and *trauma* are central and important issues in what follows:

> The end result of the analysis of transference and counter-transference may very well be the establishment of a kind, dispassionate atmosphere, such as may well have existed in pre-traumatic times. [p. 27]

In a mutual analysis, one finds

> that common feature which repeats itself in every case of infantile trauma? And is the discovery or perception of this the condition for understanding and for the flood of healing compassion? [p. 15]

Ferenczi's concept of trauma complements that of Freud. While the latter concentrated on the investigation of interpsychic processes, Ferenczi put the relationship of the individual with his environmental reality at the forefront and examined how the organism reacts to changed external conditions—whether in philogenetic speculation, drawn from the theory of genitality ("Thalassa: A Theory of Genitality", 1924 [268]), or in his questions about the relationships between the adult and the child, between analyst and analysand.

Ferenczi considered the traumatic event and its working through in therapy *from the viewpoint of man as an essentially social being* (Ferenczi wrote to Groddeck: "Psychoanalysis is . . . a social phenomenon"—Ferenczi & Groddeck, 1982, p. 45): before the trauma, there was an atmosphere of trust between the individual (the child) and his social environment (the adults)—the first phase—which is then destroyed by an extreme rise of tension in the relationship—second phase. The child seeks help from those very people who caused the rise in the affective temperature of the relationship. If help is not forthcoming—third phase—a split occurs, in which one part of the personality endures the intolerable state while the other looks on unemotionally from a distance, and offers comfort, tries, in fact, to perform the function of "an auxiliary ego", which the external world should have provided. The consequence of trauma is a permanent disturbance in the contact with social reality; the ego, the "outer layer" (Freud, 1940a) of the psychic organization has withdrawn so far "within" that it can no longer guarantee *its function of interchange.*

In the therapy, Ferenczi tried to revive the traumatic sequence and to find a new outcome by offering what had not been offered in childhood: an atmosphere of trust—in Balint's words, "*arglos*" (Balint, 1933, p. 157). He hoped that this would enable the analysand to heal the rift in his personality. Such treatment demands a specific kind of listening and sensitivity on the part of the analyst.

He wrote (Ferenczi, 1985 [1932]):

> But if the patient notices that I feel real compassion for her and that I am eagerly determined to search for the causes of her suffering, she then suddenly not only becomes capable of giving a dramatic account of the events but also can talk to me about them. The congenial atmosphere thus enables her to project the traumata into the past and communicate them as memories. A contrast to the environment surrounding the traumatic situation—that is, sympathy, trust—mutual trust—must first be created before a new footing can be established: memory instead of repetition. Free association by itself, without these new foundations for an atmosphere of trust, will thus bring no real healing. The doctor must really be involved in the case, heart and soul, or honestly admit it when he is not, in total contrast with the behaviour of adults toward children. [pp. 169–170]

The analyst's involvement, the countertransference, is thus no longer simply an obstacle; it becomes an *important instrument*—a fact that Freud had recognized already in 1912 (Freud, 1912b)—rather in the sense of St. Augustine's "fortunate sins". "One could almost say that the more weaknesses an analyst has, which lead to greater or lesser mistakes and errors but which are then uncovered and treated in the course of mutual analysis, the more likely the analysis is to rest on profound and realistic foundations" (Ferenczi, 1985 [1932], p. 15). The analyst's "strength" is thus a function of how he deals with his "weaknesses". Clarification of the countertransference is important for the analyst as well; he need only admit it: "In one case the communication of the content of psyche developed into a form of mutual analysis, from which I, the analyst, derived much profit." Ferenczi was quick to recognize the "dilemma" of the mutual analysis (p. 28), *its dangers and its limitations*, which restrict its use (p. 34): this mutual analysis can be done ". . . in so far as a) the patient needs it, and b) is capable of it in a given situation". He also refers to this mutual analysis as "only a last resort!" (p. 115).

> One has a vision of the successful end of an analysis, which would be quite similar to the parting of two happy companions who after years of hard work together have become friends, but who must realize without any tragic scenes that life does not consist solely of school friendships, and that each must go on developing according to his own plans for the future. This is how the happy outcome of the parent–child relationship might be imagined. [p. 37]

> Just as Freud's strength lies in firmness of education, so mine lies in the depth of the relaxation technique. [p. 62]

This led Ferenczi to regard Freud's position as an intellectual and impersonal one, despite his recognition of the primacy of affect in the transference (p. 54), to recognizing the *importance* of the analyst's own analysis and to envisaging it thus:

> 5. Mutual analyses: only a last resort! Proper analysis by a stranger, without any obligation, would be better. 6. The best analyst is a patient who has been cured. Other pupils must be first made ill, then cured and made aware. 7. Doubts about *supervised analyses*: last resorts: recognition and admission of one's own difficulties and weaknesses. Strictly supervised by the patients! No attempts to defend oneself. [p. 115]

He felt, however, that analysands are unfortunately often "used instead of being allowed to develop". Looking back, Ferenczi described the path he had taken. There was an assertion, attributed to Freud (Ferenczi, 1985 [1932], that

> neurotics are a rabble, good only to support us financially and to allow us to learn from their cases: psychoanalysis as a therapy may be worthless. This was the point where I refused to follow him. Against his will I began to deal openly with questions of technique. I refused to abuse the patients' trust in this way, and neither did I share his idea that therapy was worthless. I believed rather that therapy was good, but perhaps we were still deficient, and I began to look for our errors. In this search I took several false steps; I went too far with Rank, because on one point (the trans-

ference situation) he dazzled me with his new insight. I
tried to pursue the Freudian technique of frustration hon-
estly and sincerely to the end (active therapy). Following its
failure I tried permissiveness and relaxation, again an ex-
aggeration. In the wake of these two defeats, I am working
humanely and naturally, with benevolence, and free from
personal prejudices, on the acquisition of knowledge that
will allow me to help. [p. 186]

That was his plan.

The ideas he developed from this procedure are remarkable.
For example, his criticism of the sacrifice of "woman's inter-
ests", which Freud discussed, sounds like radical feminism
(Ferenczi, 1985 [1932]):

Take, for example, the castration theory in femininity.
Freud thinks that the clitoris is developed and functions
earlier than the vagina; in other words, that the little girl is
born with the feeling that she has a penis. It is only later
that she learns to give up this, along with the mother, and
to adapt to vaginal and uterine femininity. He neglects the
other possibility, that the heterosexual drive orientation is
well developed early on (perhaps only in fantasy), and that
masculinity may only take its place for traumatic reasons
(primal scene), as a hysterical symptom. It is possible that
the author has a personal repugnance towards the sponta-
neous sexuality of the woman: idealization of the mother.
He recoils at the prospect of accepting the fact of having a
sexually demanding mother and of having to satisfy her. At
one particular time he was faced with just such a prospect
by the impassioned nature of his mother. (The primal scene
may have made him relatively impotent.) *The castration of
the father, the powerful one, as a reaction to humiliation, led
to the construction of a theory in which the father castrates
the son* and, furthermore, is subsequently adored by the
son as a god. Freud plays only the role of the castrating
god; he wants none of the traumatic moment of his own
castration in childhood. He is the only one who does not
need analysis. [pp. 257–258; italics in the original]

These are hard words, indeed, and for many undoubtedly
quite shocking, but perhaps also words of arousal to take up
this attitude, this *internal* psychoanalytic position for which

Ferenczi pleaded, rather than remaining in a projective position thinking that stated truths as such—indeed, immutable
dogma—can ensure the future of psychoanalysis, or even *our*
personal future. Psychoanalysis must constantly question and
reconstruct, or else it risks losing its function and ceasing to
exist.

Ferenczi had a remarkably enquiring mind; he described
himself as "a restless mind" (Ferenczi, 1933 [294], p. 126). He
did not pretend to have "solved" problems, or to have given an
answer to the questions he raised. His last work, the *Clinical
Diary* (Ferenczi, 1985 [1932]), is a return to psychoanalysis in
the true sense of the word: a return to self-exploration. His
"experiments", as he called them, opened new paths, leading to
no definite conclusions. He would not even have recommended
"mutual analysis" (pp. 28, 43, 115) for every one of his patients, and he was perfectly aware of its weaknesses. The direction he took, however, has proved a very fruitful one. It can be
said rather simplistically that while Freud concentrated on
theory, Ferenczi's greatest preoccupation was with the expansion of the technique and practice of psychoanalysis. Freud
tended to see the patient as an object of rational study, seeking
new "insights" for his model of the mind; Ferenczi considered
his analysand as a person who was suffering and who *inter-
acted* with the analyst and *affected* him. His was the exploration of affective communication in psychoanalysis. Through his
genius, Freud constructed a "one-person" psychology. He developed the "object-related" method with very high levels of
work on the transference and countertransference, interactive
empathy (with a clear recognition of the link between empathy
and countertransference in the broadest meaning of the term),
and the use of regression. From his interest in the countertransference, he focused increasingly on the analyst as a whole
person. Every analyst who regards the analysis as an interaction implying a high degree of personal involvement and an
explicit awareness of it is heir to the pioneering work of Sándor
Ferenczi. Recognition of our intellectual heritage makes us
aware of important problems of psychoanalytic practice and
their reflection (or absence) in our theoretical frame of reference.

A number of facts have contributed to the historic fact that the work of this pioneer and the discussion that followed—which was of the greatest importance for psychoanalysis—have fallen into oblivion. These factors may be related in part at least to Freud, to his age and his personal limitations, but they owe even more to the relationships between analysts and their *fears* and *resistances* about questioning or even abandoning positions regarded as safe and sure. As Balint says (1968a, p. 207), "The discord between Freud and Ferenczi had a traumatic effect on the analytic world". It created a *silence*, which suppressed the whole controversial affair as being disturbing and "dangerous". After Ferenczi's death, analysts, impressed by the fact that even so close a friendship could be seriously troubled by such problems, became extremely prudent in their discussions about technique, even though they all admitted that the analysis of transference was an essential theme. Strachey (1934), with his notion of the mutative interpretation in the transference, is quite clearly following Ferenczi, as is Sterba (1934) with the importance of identification.

The problems of regression and especially of countertransference seemed to disappear from discussion until Alice and Michael Balint published their article on countertransference in 1939 (Balint & Balint, 1939). It was Michael Balint, the literary executor of Ferenczi's will, who drew the attention of the British Middle Group of independent thinkers to this problem (Kohon, 1986); Donald Winnicott (1949), Paula Heimann (1950), and later on many others took it up. In England, a group has existed which for many years has been sensitive to these problems, perhaps in part under the influence of Melanie Klein, a pupil of Ferenczi, and in part through the influence of others, for example Ella Sharpe (1930). This tradition—which considers the importance of the "setting" (Winnicott, 1956), and of other concepts such as "framework"—opens new perspectives.

Refocusing our history on the facts that have punctuated it seems fundamental to me and could serve to liberate us from a certain sectarianism, and indeed from the too great and therefore unanalysable influence of "sorcerers" encountered during training. This might well restore a historical perspective to the remarkable adventure of investigation of the internal life. It is an extraordinary enterprise, started by a *genius* and continued by *prodigious* personalities, one of whom was distinguished, not only for his great charm and generosity, but also for his *intellectual courage*, independence, and uncommon honesty— Dr. Sándor Ferenczi.

Resumption of the controversy

Fears have been expressed, especially latterly, in the wake of a sort of "rediscovery" of Ferenczi in France, that towards the end of his life he was at the root of a "re-establishment" of the thesis of the innocence of the child, perverted and made neurotic only through the behaviour of adults. These fears reveal a misunderstanding of the complexity of a man who, like Donald Winnicott, attributed a fundamental importance to the human *environment* of the child, but not without acknowledging the child's drive apparatus or his "Oedipus complex". After all, it was he who thought that the Oedipus complex is the nuclear complex and that a man's attitude towards it determines a normal person's character (1912 [92], p. 21). His is not the position of an external environmentalist. The whole of his work and his correspondence with Freud are proof to the contrary. An *interactional* model does not imply reversal of the generational order, as Grunberger (1974) thinks. The flexibility with which he examines the interpenetration of psychosexual states is in line with Freud's "*après-coup*" ("deferred effect") and in no way assumes abolition of the theory of those stages. No one doubts that Ferenczi puts less emphasis on the anal-sadistic stage than does Grunberger, but this is no reason to treat him as a dissident by a belated excommunication (Grunberger, 1974). Grunberger does recognize that Ferenczi's merits lay in considering the analysand's total personality, not only his symptoms; his criticism of Ferenczi with regard to the centrality of the Oedipus complex is a matter of personal dislike.

Cahn (1983) draws attention to Ferenczi's importance in the theories concerning the psychoanalytic technique, and especially its possible *traumatizing* potential, the range of countertransference movements that can be manifested in the breakdown of *holding* functions ("analytic framework"), and the fundamental importance of recognizing and analysing them. Donnet (1983) attributes to Ferenczi a primary role in relation to *repetition* in the treatment. Donnet feels that Freud was never totally reconciled to repetition, especially if in treatment it took the place of remembering. In fact, he assigns to the repetition compulsion a position alongside the death instincts, contrary to his earlier viewpoint and Ferenczi's that repetition could be a *pleasure* and may pave the way to a better *mastery* of repeated contents.

One should not necessarily follow Donnet's distinction between an interpretation that emerges in a session and a construction based on pre-established knowledge, because, in the end, this so called "pre-established" knowledge is part of what the analyst has stored away in his memory, and which emerges at the time he formulates his interpretation or his construction. Nevertheless, to stipulate that Freud's inclination for constructions can be considered as the expression of his temperament and his recourse to pre-established knowledge, whereas Ferenczi's way of working is based more on introjective and projective processes and on formulations freshly acquired from the material, does seem to provide an adequate description of the two positions. In this sense, Melanie Klein can be seen to be a follower of Ferenczi's technique—although hers was probably less gratifying for the analysand, and her use of pre-established schemas less empathic. In France, one of the reasons for a certain amount of reserve towards Ferenczi is due to this descendancy: Sokolnicka, a pupil of Ferenczi, was the analyst of Pichon and Laforgue—a fact that was not necessarily a very good reference for the post-war generation (Roudinesco, 1982, pp. 288, 306; 1986, pp. 175–177). But, in my opinion, this is not sufficient reason for continuing either to ignore him or to indulge in criticism without sufficient knowledge of his work.

While Ferenczi (1928 [282], pp. 84) considered that the training analysis should be taken as far as possible, Freud, already in 1937 (1937c, p. 248), thought that it could be reduced to giving a relatively brief experience of the unconscious. Girard compares Eitingon and Ferenczi—in other words, the Berlin and Budapest schools. In this historic comparison,

Eitingon favours the institutionalization and integration of technique through a method that could lead "to the obessionalizing of the psychoanalytic training" in order to give a sense of security. Ferenczi, in his inner freedom, tried to develop a technique in which there was no fear of an "affective trance", and he emphasized the development of the personality (Girard, 1984):

> Each has a grandiose vision about the transmission of psychoanalysis anchored in the pursuit of his original demand from Freud—in Eitingon's case to be taught, in Ferenczi's case to be cured, two permanent modes of entry into the process of analytic training. They also form two complementary modes in the organization of the training since the Ferenczian perspective largely dominates the courses, both on the formal level, with regard to duration and on the ideal level, and with regard to the results. [p. 120]

CHAPTER THIRTEEN

An example: narcissism or the history of a concept

> My son David, aged 3, looking at himself in the mirror: "This mirror isn't 'wight'! . . ."

The development of concepts in the history of psycho-analytic ideas merits exemplification. I shall take just one concept to demonstrate—narcissism. It should be remembered that Freud himself was aware of the to-and-fro between experience and concept. In order to answer questions as soon as they are posed in experience, new concepts have been adumbrated in attempts to cover phenomena appearing progressively in different perspectives. In the course of re-elaboration, emphasis has constantly changed through modification, looking back, and nuances of meaning. Freud described this as follows (1915c):

> Physics furnishes an excellent illustration of the way in which even "basic concepts" that have been established in the form of definitions are constantly being altered in their content. A conventional basic concept of this kind, which at

213

the moment is still somewhat obscure but which is indispensable to us . . . is that of *instinct.* [p. 117]

This to-and-fro caused imperceptible changes of meaning and generalizations so as to make "falsification", in Popper's sense, difficult—that is, testing the validity of psychoanalytic theory. Despite certain efforts (e.g. Bolland & Sandler, 1965), this is still the case today.

A continuous questioning

If psychoanalytic theory is not unequivocal, there are many reasons for it. Freud did not create a closed system in which concepts would have had a distinct meaning once and for all, a network, a scaffolding in which it would be imprisoned. In the course of his development, he sought to define both his own and his analysand's inner psychic realities in such a way that he could perceive and understand them through successive and alternating identificatory movements. Thus one witnesses in his work, and subsequently in the work of other analysts, a continuous questioning. It is convincing enough to look into the history of a concept.

The term "narcissism" was coined by Näcke in 1888, in connection with Havelock Ellis' ideas, and was taken up by Sadger in the Viennese psychoanalytic Circle in 1908 (Nunberg & Federn, 1962). It was used two years later by Freud in a note on the *Three Essays on the Theory of Sexuality* (1905d) and later in the study on Leonardo da Vinci (1910c). Regarding Leonardo's homosexuality, Freud calls to mind the love of self— but also the love of the image of the small child that we all were, i.e. the love of a double (mother, twin) with whom we were merged, an image to which we can return if we are disappointed with those closest to us in our family circle. Such are the "narcissistic movements" constituting the problem of narcissism as Freud presented it.

Narcissism, a *complex concept,* appears, therefore, as a *nodal* point, the outcome of several strands of thinking. For psychoanalytic reflection, it is an idea of utmost importance, and its contradictions are probably those of the mind itself, deriving from the difficulty of defining the latter's functioning.

This multi-faceted concept reflects the way the subject *sees himself*, his image, his double. With Freud, it is clear that we have to resolve several difficulties on the path of understanding the human mind in its complexity, with its conscious and unconscious dimensions and its links with the instincts, particularly the sexual instinct.

Turning in on the self, the search for the self *in the Other—* there are so many forms of narcissism: there is the everyday narcissism accompanying our perception of ourselves, and *"Selbstgefühl"* [self-esteem], the sense of the self, of Freud (1914c). Let us bear in mind that this form of self-love is a basic form of love. The commandment "Love thy neighbour as thyself" obviously implies that one loved oneself. At the one extreme, there is the turning in on the self in restorative sleep; at the other extreme, there are the various forms of self-love. One could certainly contrast love of the self with love of the Other (the love object). In this view, narcissism is the opposite of object cathexis. And yet, even if such an extreme can be imagined, these two extremes are not necessarily opposed in adult relationships; rather, they accompany each other.

In 1910, in "The Psycho-Analytic View of Psychogenic Disturbance of Vision" (1910i) Freud refers to ego instincts, having already previously alluded to "ambitious ego drives" (1907b, 1908c, 1908e, 1909c). His theoretical concern, above all in his controversy with Jung, was to maintain his concept of sexual libido, which, apart from the object side, also has a narcissistic side.

If the concept of *primary narcissism* is understood as "an early state in which the child cathects its own self with the whole of its libido" (Laplanche & Pontalis, 1967, p. 337), *secondary narcissism* would be the internalization of a relationship—especially that with the mother. It would constitute one of the basic essentials of the mind as can be seen in states resulting from *inadequacy* or deprivation of such love (Bowlby, Spitz, and others). Interpretation of what Freud called "*primary narcissism*" raises a series of questions. Is it really one stage among others, and ought psychoanalysis to see the child's development as a succession of distinct stages, among which

primary narcissism would find its place? Or should it be considered as "a phase, or formative moments, marked by the
emergence of a first adumbration of the Ego and its immediate
libidinal cathexis" (Laplanche & Pontalis, 1967, p. 338)? Does
the hypothesis of such a formative moment contradict, for example, Michael Balint's idea of "*primary love*" (1965)? Primary
love could be conceived of as being close to narcissism: "Love of
the object is a transitive fraction, where alternatively the object
is either the mother or child. The child becomes the object of
the object in the relation of illusion of the mother child unity"
(Green, 1979). The development of the human being of necessity involves oscillations between narcissistic love and libidinal love. Secondary narcissism, as it was described by Freud
and since studied by others (including Luquet, 1962), would
represent the mother's love introjected by the child, who, once
separated from her and freed from symbiosis, aware of his
independence and otherness, would love himself in the same
way his mother loved him. In other words he could only love
himself ("narcissistically") as he was loved ("libidinally").

"Love of the self" ["*Selbstliebe*"] (Freud, 1915c) in a system
of turning in on the self may involve pleasure in the erogenous
zones: these auto-erotic pleasures are not to be confused, however, with the concept of narcissism. The latter is wider and
related to a different dimension from auto-erotism, which covers only a small sector.

. . . and the death instinct

For the ego, from the start of life, while reality does not yet
include the polarities *internal* and *external,* subject–object,
self–others, narcissism is the cathexis of vague unstructured
perceptions. Without the intervention of secondary processes,
it implies a disregard of reality, which may have consequences
as extreme as self-destruction. The "*death instinct*" can be
understood as a tendency to return to a narcissistic system
with that non-structuring that characterizes it. It can also
be like the psychic representative of the law of entropy,
in a battle against that anti-entropy which is life. If the aggres-

sion contained in the death instinct puts the Self in danger, it is also because the distinction between self and others is ignored.

André Green (1967) notes that in Freud's thinking, the aspiration to a state of total unexcitability is a constant. This idea implies that every external stimulus, but also every change, poses a problem. What "weight", then, does the concept carry? Is narcissism the ally of what Freud later called the death drive [*Todestrieb*] (Freud, 1920g), forcing the individual to close in on himself—a zero degree of excitation—death? Or is it also, in another context, in terms of "narcissistic provisions", the necessary accompaniment to all our acts, as I said on the subject of secondary narcissism?

A remark attributed to Béla Grunberger is noteworthy: namely, that in order not to be depressed, considering the brevity and precariousness of existence, one has to be manic—or, rather, *narcissistic*, in the sense of the protective narcissism of life.

This preponderance of narcissism at the very beginning of life has been compared with cognitive egocentricity as highlighted in Piaget's research; so cognitive decentration would be paralleled with an affective "decentration", and the dawn of a symbiotic world with a narcissistic charge would open into a perpetual quest for an equilibrium always to be reconstructed between cathexis of the self and cathexis of the relationship with others.

. . . and erotism

Parallel to the importance Freud placed on sexuality, there is an essential *narcissistic* cathexis of the sexual organs and in particular of the visible sexual organ, the phallus (an idea already studied in his time by Jenö Harnik, 1923). Other pioneers of psychoanalysis have also emphasized cathexis of parental functions (Reik).

In retracing Freud's antagonist model, Grunberger (1971), redevelops the dynamics of this major antagonism, as one of the narcissistic and libidinal contributions (drives such as anality and orality), which come together in the Oedipus complex. From the original narcissism, Pasche even postulates the attraction of objects snatching us from a life under a narcissistic regime (an *anti-narcissistic* principle).

In 1913, Ferenczi, probably the first psychoanalyst to have conceived "a developmental line" for a specific area of psychic functioning, wrote as follows (1913 [1911]):

> Auto-eroticism and narcissism are thus the *omnipotence stages of erotism*, and, since *narcissism never comes to an end at all, but always remains by the side of object-erotism*, it can thus be said that—in so far as we confine ourselves to self-love—in the matter of love we can retain the illusion of omnipotence throughout life. [p. 234]

Under Sándor Ferenczi's influence, Grunberger roots narcissism firmly in prenatal life. The memories or reconstructions of a state of bliss free from conflicts and excitations ("Paradise lost", or "the Golden Age") would be responsible for the yearning of mankind to find this happiness again. Like Ferenczi, Grunberger interprets the incestuous wish as being linked to nostalgia for this return into the mother's womb. The Oedipus complex would thus be the outcome of both narcissistic and instinctual wishes. Then, the incest taboo—considered by Lévi-Strauss to be the nucleus of civilization—would primarily play the role of protection against the narcissistic wound: parental prohibition would in effect mask the child's impotence. In reality, his physiological incapacity and his immaturity would prevent his finding the bliss of prenatal life through incest, but he attributes it to parental prohibition.

The loss of omnipotence leaves a profound effect on man—an idea shared by Ferenczi's followers. Róheim, for example, refers to the neoteny of man (the child, in his total helplessness, becomes the infant–king through his mother's care). Moreover, Grunberger considers all the manifestations of civilization as representing various attempts by man to bring about a re-establishment of narcissism.

When I say that the balance between narcissism and the object relation has always to be recaptured, I am implicitly alluding to imbalances in narcissism, a subject that preoccupied Freud.

Narcissistic pathology would, in a first approximation, include an inflated narcissism at the expense of the object relationship. The nomenclature may change, but the problem remains the same—viz., what role will narcissism play? Freud talked about "narcissistic neurosis" differently from the way it is used today. For him the term corresponded to what today's psychiatrists and psychoanalysts call "the psychoses". Contemporary writers, on the other hand, especially Kohut and Kernberg, put the "narcissistic neuroses" in-between the psychoses and the so-called classical neuroses, and they more or less include "borderline states".

Gaping/mirror

For Kernberg and for the Kleinians, envy and oral–sadistic rage are the crucial determinants of character problems akin to borderline states from which they are differentiated mainly by their recourse to primitive mechanisms such as splitting, denial, projective identification, pathological idealization, and a system of omnipotence. For Kohut, on the other hand, narcissism follows a line of development, and its pathology would therefore result from an arrest in that development at the level of the grandiose self, of idealized parental images, of clinical phenomena such as transference onto the twin, etc. The work in the United States of these two schools of thought (well after Grunberger's work in France) has brought out the importance of *narcissistic problems* which today's psychoanalysts would encounter more frequently than before. I shall return, later, to the problem implied in this assertion.

But, as I said, this concept was destined, at a certain moment, to explain the phenomena that confronted Freud and which he was aware of because of Jung's and Abraham's work at the Burghölzli clinic in Zurich. Jung was preoccupied by the problem of psychoses, and it was at his request that Freud commented on President Schreber's book. In 1908, Abraham, for his part, paved the way for the understanding of psychosis: "The psychosexual characteristic of dementia praecox is the patient's return to auto-erotism."

So the meaning attributed to "narcissism" follows its historical course. At first, in Havelock Ellis and Näcke, it is a perversion in which the subject's sexual object is his own body: the
love of oneself, in psychosis to the total exclusion of the rest of
the world, leading to the elaboration of the concepts of primary
and secondary narcissism (the latter encompassing every contribution through introjection of "the object") that I have already
mentioned. Alongside "healthy" everyday narcissism there is
defensive narcissism, which may lead to the negation of the lost
object, at the risk, to the subject, of leaving the real world and
being precipitated into madness. Assuming that the narcissistic
subject is preoccupied only with himself, Freud thought that he
is not likely to make a transference. It is a debatable hypothesis
for the majority of today's psychoanalysts, since we talk about
"narcissistic transference" (cf. Kohut), or at least about narcissistic elements of the transference.

The narcissistic love of self, through a speculative relationship mirrored in another person identical to us (our twin) or
one who is supposed to be like us—such as the mother at a
certain time of life (the child loves himself in the light of her
eyes) or the love of someone resembling the self in whom we
find character traits of our own—these are the origin of Freud's
idea that some homosexual attachments are rooted in this form
of narcissistic love (1910c).

On Freudian lines, Federn (1952) devoted himself to studying the narcissistic cathexis of the *frontiers of the ego*. These
would be disturbed in different degrees of states of depperson-
alization and in psychoses in which, following Nunberg, he emphasizes the importance of the *loss* of the object. Feelings of
strangeness and depersonalization reflect problems in experiencing the external world, and, at the same time, the image of
the body ego, the feeling of the reality of objects depending on
narcissistic libidinal cathexis of Freud's *"Selbstgefühl"* (1914c).
According to Federn, it could be said that knowing the object,
knowing its limits and its permanence, depends on libidinal
cathexes of the ego: "Here narcissism is based on the object",
as Lebovici (1960) says—a theoretical precedence in any case,
and perhaps a factual one in the first stages of human life.

Narcissism follows us like a shadow. It refers to the concept
of an original Paradise (perhaps intra-uterine), and then to a

stage in which the child believes in the omnipotence of his thoughts (Freud, 1912–13), whereas he is not obliged to be preoccupied with the satisfying of his hopes and ignores desire through his lack of an individual existence. After this first stage of bliss comes a reality in which the little man feels frail and impoverished. Freud talks about the child's helplessness, which stamps the narcissistic vulnerability of the human being and the points of later possibly narcissistic regressions. This injury may be sufficiently healed by narcissistic *supplies* in the mother–child relationship and in later development. Otherwise it leaves a gap that is never filled, a constant vulnerability and a perpetual thirst, to be recognized and confirmed in his narcissism.

One could imagine, if one brings together all these conceptions into a diachronic viewpoint, transformations of narcissism at every developmental stage, marked by the organization of self-representation: the time when this latter is "cathected" with narcissistic libido is that fundamental experience named by Lacan as the "*mirror stage*" (the "jubilant assumption" before the mirror being the expression of the affect before the narcissistic image of the body—Green, 1973). For this cathexis, Winnicott emphasized the role of the mother, her reactions, and the introjection of her significant attitude.

The mirror stage is in some way the paradigm of a dual relationship; for Lacan (1966) it is seeing the image of the self in the mirror as if it were another. The importance of the name only adds a triangular level, through the law. According to Lacan, therefore, it is not a problem of either real or symbolic existence. It is the image that matters: to see oneself in the Other. So the ego would be a construction, an imaginary statue, a mould into which one's abandoned identity is thrown.

For Lacan, who in 1936 pointed out the importance of the discovery of the body no longer perceived as a part body but as a unity (Mannoni, 1979); the child's fascination with the image of the Other constitutes an anticipation, by identifying with that image, of a bodily unity, which will be attained only later. The central idea is that this vital experience makes the *ego* an imaginary structure. The mirror stage constitutes the source of later identifications.

In this respect, Lacan is in agreement with many authors who think that the narcissistic cathexis by objects and cathexis of object libido are not two separate identities, but coexistent modes. The Other, in the transference, is felt as being like—not identical to—the absolute master (in Hegel's sense of the master and the slave) or the absolute slave, feigning death through his silence. This bears some resemblance to Rank's (1914) idea, according to which the experience of the double is a confrontation with the Other—with the counterpart, the twin—and, in the end, an encounter with one's own death. From this viewpoint, the dual relationship is not so different from the relationship with the self. They have close ties, a fundamental fact for the understanding of the narcissistic dimension.

Nevertheless, there is some convergence in the apprehension of this difficult concept that psychoanalysis could not do without. The fact that so many authors devote themselves to it and attempt to distinguish it, to relocate it in their conceptual systems, shows how necessary it is. Even if the emphases are differently distributed, there are surprising agreements. Lacan, for example, is akin to Grunberger in his view of the existence of an important "pivotal" point between narcissism and the Oedipal structure. In the Freudian tradition, it is the point at which the Oedipal situation would fail through its impossibility, and thenceforth narcissistic libido would triumph over object libido and the subject would renounce the object in order to survive.

Sublimation, culture

We cannot talk about narcissism without mentioning the *ideal.* As I do not wish to enter into the problems of the ego ideal (Chasseguet-Smirgel, 1975), I shall restrict myself to stressing that the ego must always be seen in relation to its ideal, which means that it takes sides with some objects against others and that it also includes the ideal that has been assigned to it (Freud, 1916–17), this always being narcissistic. The subject's equilibrium is always threatened if he has the feeling of not matching up to the ideal. However, it is an asymptotic rapprochement, the ideals being what are aimed at but never absolutely achieved. Everything, in the Freudian topographical system, takes place *within* the ego, the superego being only a "*Stufe im Ich*", "a step on the ladder", a "degree".

I have alluded only briefly to the important links between narcissism and creativity, between narcissism and sublimation. Less a theoretical question, it is a matter of knowing whether sublimation is "effected through the intermediary of the ego, transforming sexual libido directed towards the object into narcissistic libido with different aims" (Freud, 1923b, p. 30). To what extent, in cultural activity, are our narcissistic ideals involved as a major driving force? To what extent is there a link between the culture and the narcissistic ideals of our childhood? In *Civilization and its Discontents* (1930a) Freud raises the question of the fate of civilization and culture in terms—borrowed from mythology—of Eros and Thanatos. He seems, through new conceptions, to be seeking to compare the aims that Eros offers us as historic ideals through civilization (traditional civilizing ideals) with "*ananké*", necessity, and Thanatos, which tend towards zero, towards destruction. The problems of narcissism and sublimation may throw light on the requirements of our civilization through ideas that Freud developed after 1920.

Narcissism has become, in some sense, a "fashionable" concept in the psychoanalytic literature. It is as if, after sexual "liberation", there could be some awareness that all problems have not been finally resolved, as if the questioning had been displaced from sexual libido or the aggressive drive to narcissism. Is it really to do with readjusting conflicts according to cultural, educational, and socio-economic factors—an internal shift of the eternal conflictive forces in man—or with the change of perspective in which it is viewed? If the latter, then why? It is difficult to say. There are several indications that an education emphasizing prohibitions on sexuality favours the appearance later on in adulthood of psychic conflicts in the superego, while other educational and cultural models lead to other traumata, such as over-gratification (in the sense of "frustrating the frustrations"), abandonment, disinterest, non-encounters.

It may equally be supposed that an element of disappointment with regard to shared cultural ideals, a greater isolation, a turning in on the self would favour the creation of a "narcissistic culture" (Lasch, 1978). The sociologist Richard Sennett (1974) thus sees a link between "the fall of public man" and "the rise of private man", just as, after the death of Augustus, the decline of public life resulted in an unprecedented quest for religious transcendency. So, in us, a disinterest in all things public might well provoke the tendency to turn in on the self.

Perhaps there are certain times in a culture that favour man's self-questioning. Others, however, think that pathology always remains the same, as do the configurations of the personality, and that which changes is the style of questioning, seeking to discern through multiple shifts what in the end remains inapprehensible and unexplained. Only gradually does the unconscious disclose its secrets, although some ways of grasping it can result in more-or-less limited clarification. It is not so much a matter of requestioning the great discoveries of psychoanalysis, but of emphasizing that there is assuredly no definitive reply to all the questions of existence. Narcissism, however, is undoubtedly one of the important concepts that help us to understand man in his eternal quest for happiness and the proximity of the happiness to death.

Narcissus, lover of himself, was destroyed by that love—a myth whose profound meaning is inexhaustible and to be contemplated again and again.

EPILOGUE

Turning and turning in the widening gyre
The falcon cannot hear the falconer;
Things fall apart; the centre cannot hold;
Mere anarchy is loosed upon the world
. . .
The best lack all conviction, while the worst
Are full of passionate intensity.

Yeats, "The Second Coming"

The Freudian image of man

I t is more than fifty years since Freud died, on 23 September 1939. His family doctor, Max Schur (1972), writes as follows:

> While I was sitting at his bedside, Freud took my hand and said to me: "My dear Schur, you certainly remember our first talk. You promised me then not to forsake me when my time comes. Now it's nothing but torture and makes no sense anymore." I indicated that I had not forgotten my promise. He sighed with relief, held my hand for a moment longer, and said: "I thank you", and after a moment of hesitation he added: "Tell Anna about this." All this was said without a trace of emotionality or self-pity, and with full consciousness of reality.
>
> I informed Anna of our conversation, as Freud had asked. When he was again in agony, I gave him a hypodermic of two centigrams of morphine. He soon felt relief and fell into a peaceful sleep. [p. 529]

Who was this man? What had he understood of the human being throughout his long life, the course of which passed from the Emperor Franz Joseph to the Austrian republic and the "*Anschluss*"? What is the significance of his image of man for our culture and for today's generations?

The image he constructed of man has often been described as *pessimistic*. Myself, I would prefer to call it "realistic": it has unfortunately been too frequently confirmed by contemporary history and the general history of humanity. He challenged illusions—the illusions of his world, of the bourgeois society of the nineteenth century, and the *Utopian* thinking of the twentieth century and the dreams that followed the two World Wars. His endless wish to know what, fundamentally, man was turned into a sort of *philosophy of suspicion*—suspicion that man is not all that he imagines—while at the same time knowing that questioning the Other or the self immediately awakens an automatic antagonistic force, the resistance. This questioning and suspicion on the one hand and the *resistance* emanating from it on the other are basic elements of psychoanalytic science, a science that, resulting from practice, also contains the facts brought to light by that practice as an awareness of the processes that develop in it. Psychoanalysis is a practice and a theory—a *body of doctrine*.

As Stefan Zweig said, nobody is a great man in the eyes of his valet. In Vienna, where little stories and the passionate atmosphere of dissensions between the first analysts are well known, not enough notice was taken of the great moment of cultural history that Freud's work represents. In fact, the trivial stories did nothing to deprive either the man or *his work* of grandeur.

Psychoanalysis is a search for the truth of man about himself: an ascetic undertaking, which goes not without injury to his self-esteem, to what is called his "*narcissism*". Perhaps this is Freud's greatest achievement in cultural history: to have enabled the European and American cultured man of the twentieth century to see himself reflected in a mirror. Psychoanalysis adds its own way of understanding and interpreting to everything that biology, anthropology, linguistics, and the whole of so-called human sciences have said about man. What appears, therefore, is an image that is not nurtured by an *a priori philosophy*, but by clinical experience. Freud's antipathy towards philosophy as a *gratuitous act* is well known. Whatever he says, he endeavours to verify by himself in the smallest details of his patients' associations. This microscopic work on the mental life of man is a unique opening in the history of humanity. Listening to the free associations of the Other—as free as is possible—is a *unique* and very special *experience*.

Wollheim (1979) expresses it well when he says:

> Freud's theory shows man to be endowed with a very complex internal structure. This internal structure changes. It matures, and also it is modified by experience which can be both of outer and of inner reality. But if experience modifies structure, structure mediates experience. It determines how man reacts to experience, and this reaction, like the experience it reacts to, can be either external or internal. Structure, experience, reaction—Freudian theory shows these to be interdependent, and yet capable of being independently studied. [p. 6]

Man in conflict

Freud postulates that *frustration* plays a large part in the genesis of psychic disturbances. At first, he was thinking particularly of so-called sexual frustrations, and one might have hoped that the decrease of restraints in that area would have had a prophylactic effect. Yet this has not been the case at all. It shows that alongside environmental and educational factors, the influence of the passage from omnipotence/omniscience to an awareness of a *limited* human *condition* is of paramount importance: the fact that some of our actions expose us to dangers, that satisfaction is not automatic and limitless, marks the passage from the *pleasure principle*, from self-satisfaction, to the *reality principle*. The latter may, according to the social conditions, involve greater or lesser constraints, but it always implies *limits* (Freud, 1911b).

Foucault (1967) recalls that *there is no civilization without prohibitions*, in part linked to real dangers. In a complementary balance with gratification and pleasure, frustration acts as the counterweight. This dialectic opposition is an integral part of the basic conditions of human existence and proof that man's complex development escapes the simplifications or manipulations that they will be able to inspire. Man is defined by the need to abandon the pleasure principle, the principle of self-satisfaction, for a reality principle, within the limits of which, by contrast, he will be able to create situations that enable him to procure pleasure. The reality principle appears as the *modification* of the pleasure principle by *taking reality into consideration*. To do this, the search for satisfaction abandons the shortest route and postpones pleasure according to the conditions and rules of the external world, in order, finally, to achieve a surer pleasure.

Another factor highlighted by Freud's work is the *Oedipus complex*. Far from being reduced to a concrete simple situation, it is understood as a violent and inevitable conflict between desire and prohibition—between the temptation of incest and parricide and the laws that condemn them. Structural anthropology confirms that all societies have certain forms of prohibition against incest. The failure of wishes confronted by impossibility and prohibition (the boy's outburst of love for his mother and hatred of his father) creates the Oedipus complex, the central and ubiquitous knot for understanding man. Freud's first contribution to social psychology and anthropology was *Totem and Taboo*, written in 1912–13, in which he takes the origins of religion, morals, society, and art, as well as the nucleus of the neuroses, back to the Oedipus complex. The problems of social psychology could thus be reduced to obstacles to instinctual satisfaction mostly represented by the father (or brothers, who are objects of envy and jealousy for Melanie Klein). (Obviously I am not unmindful that for Kleinians envy is primary and not a displacement: viz. primary envy of the mother's capacities felt by the impoverished helpless child at the mercy of his parents.)

The Oedipus complex thus underlies an eminently dialectic process, including at one and the same time both acceptance of the law and revolt against it, with the ambition of taking the father's place. The Oedipal structure implies the existence and prohibition of desire—desire and the problem of evil. According to Christian *myth*, the son's acceptance of castration ought to suppress the conflict; evil, after all, represents less the forbidden wish than the tension aroused by the conflict that is a testimony of temptation. In the Oedipal myth, on the contrary, it is the intimacy that calls for punishment, conflict being experienced as inherent in the fate of humanity and conferring a tragic dimension on it. Therefore, in the psychoanalytic viewpoint, it is *the inescapability of conflicts* that designates the tragic in the human condition. The tragic is not, however, necessarily pessimistic; pessimism is the expectation of the *worst solution.*

In 1897 (31 May) Freud (1950a [1897]) wrote that incest is an antisocial fact which civilization, to exist, must *gradually renounce.* The following year he declared that civilization would be responsible for the great spread of neurasthenia (1898a), with the suppression of sexuality requiring repression. The juxtaposition of these two sentences enables us to grasp the whole problem as Freud conceived it. It is true that in the *Three Essays* (1905d), Freud puts sexual repression as an interval factor in latency, which would occur without the intervention of education.

Let us remark on the ambiguity—or, if one prefers, the richness—of this analysis, in which civilization seems to be imposed from *outside* (Freud, 1908d) and at the same time linked to *biological* factors which would behave along the lines of civilization. It is nonetheless true in the majority of his work, as in *Totem and Taboo* (1912–13), that Freud is keen to examine points of tension and possibilities of synthesis between the demands of archaic drives and drives in the life of society. Later, in *Civilization and its Discontents* (1930a) and in *Why War?* (1933b, a reflection on civilization and social life), he directs his thoughts to the turning of man's destructive *aggressivity* against himself: the death instinct (Freud, 1930a):

> The element of civilization enters on the scene with the first attempt to *regulate these social relationships.* If the attempt were not made, the relationships would be subject to the arbitrary will of the individual: that is to say, the physically stronger man would decide them in the sense of his own interests and instinctual impulses. Nothing would be changed in this if this stronger man should in his turn meet someone even stronger than he. Human life in common is only made possible when a majority comes together which

is stronger than any separate individual and which remains united against all separate individuals. The power of this community is then set up as "right" in opposition to the power of the individual, which is condemned as "brute force". This replacement of the power of the individual by the power of a community constitutes the decisive step of civilization. [p. 95, italics added]

Freud thus emphasizes that civilization is built *at the expense* of individual aggressivity, on its renunciation and the consequences of it: the prohibitions.

On the basis of *directly observable clinical phenomena*, Freud stresses the role of *culture* in the *difficulties* that man encounters. His ideas are summarized in *An Outline of Psychoanalysis* (1940a [1938]). He shows that neurosis takes its point of departure in childhood, when "instinctual demands encounter *a feeble and immature Ego incapable of resistance, failing to deal with tasks which it could cope with later on with the utmost ease*". He recalls that "in the space of a few years the little primitive creature must turn into a civilized human being; he must pass through an immensely long stretch of human cultural development in an almost uncannily abbreviated form". The biological nature of man again seems to be born by "hereditary predispositions" even if "upbringing and parental influence remain important". *The Nietzschean desire for a "powerful, uninhibited ego* may seem to us intelligible", but, at the same time "it is *in the profoundest sense hostile to civilization*" (Freud, 1940a [1938], pp. 184–185).

Freud questions the importance of the work of *mourning* that is necessitated by the acquisition of civilization: above all, mourning an omnipotent megalomania and accepting the *limits* of the Self and of reality. *Violence and rage* would be the outcome of the injury inflicted by the inaccessibility of megalomanic ideals. Such *mourning* is *impossible* unless reality can be accepted. Rejection of it can only result in the creation of an increasingly narcissistic and closed world more and more detached from it. In that way, the Law is no longer an Oedipal game, implying the recognition of successive generations and the control of pleasure, but instead is experienced as being imposed by an archaic and harsh superego projected onto the external world and against which, for self-defence, a merciless battle must be waged. Behind this *rebellion* is hidden the problem of *abandonment*, of the narcissistic *wound*, and of *depression*.

Freud regards civilization as a creation based on the renunciation imposed on instinctual wishes:

> . . . *every individual is virtually an enemy of civilization* though civilization is supposed to be an object of universal human interest. . . . It includes on the one hand all the knowledge and capacity that men have acquired in order to control the forces of nature and extract its wealth for the satisfaction of human needs, and, on the other hand, all the regulations necessary in order to adjust the relations of men to one another and especially the distribution of the available wealth. [1927c, p. 6, italics added]

He takes up the theme of the balance between the instincts and culture in *The Future of an Illusion* (1927c) and in *Civilization and its Discontents* (1930a). His view of man is less than flattering (1930a):

> . . . *men are not gentle creatures who want to be loved*, and who at the most defend themselves if they are attacked; they are, on the contrary, creatures among whose instinctual endowments is to be reckoned a powerful share of aggressiveness. . . . The existence of this inclination to aggression . . . is the factor which disturbs our relations with our neighbour and which forces civilization into such a high expenditure [of energy]. In consequence of this primary mutual hostility of human beings, civilized society is perpetually threatened with disintegration. [pp. 111–112]

At best, some equilibrium is forged between frustration through the *non-satisfaction* of instincts on the one hand and, on the other, the narcissistic *satisfaction* procured through the cultural ideal, a pride in having been able to fulfil one's objectives.

The situation of infantile helplessness ["*Hilflosigkeit*"] nevertheless drives man to *movements, beliefs*, or *religions* that promise him protection while forcing his participation in the omnipotence of gods or a leader. The myths of our time are analysable in the same way as those of former times. Whatever the ideology, it can be studied with regard to its establishment and its development. Kekulé, for example, could have had personal motivations for discovering the benzene ring; knowing them and recognizing them does not in any way diminish the value of his discovery. But our narcissistic vulnerability rebels against such studies because they touch on the sacrosanct myth of objectivity. If this is true for science, it is even more true in the domain of ideology.

In Grunberger's opinion (1957), "the public is always frankly hostile or ambivalent towards psychoanalysis", because it "challenges the narcissistic defences like, for instance, an ideology, a mysticism, a religion. For we know that men are anxious to keep their 'convictions' intact and demand respect for them, defending them in the face of all that is logical". Even if they defy good sense, the convictions may be narcissistically "cathected" so well that they cannot be reached without affecting the person as a whole. The most irrational of these convictions are defended with a passionate ardour that does not bear discussion, except sometimes in an undertaking as honest and far-reaching as an analysis in which the subject—admittedly under the pressure of suffering—accepts being challenged. Grunberger again stresses that psychoanalysis contributes a good deal of narcissistic gratification since the subject is accorded time for himself and that it is in the framework of this enormous gratification that he manages to free himself from certain convictions that "lack good sense", while they are counterbalanced by a narcissistic input from the psychoanalysis. If joining Freud can be none other than diving into a strange world, a strange language—a world of ill people, a diagnostic language of formidable technicality—then that strange world is the world in which we all, in the end, are living (Norman O. Brown, 1959).

Fragility of civilization

It is cause for justifiable astonishment that, in the study of human phenomena in general and social phenomena in particular, so little use has been made of this method, which has made interpretation of the human dialogue possible and drawn concepts from it. Only after the Second World War was it asked, "*How was all this possible?*" The Frankfurt School and its adherents (Fromm, Adorno, Mitscherlich, Habermas, etc.) studied authoritarian personalities, the role of fathers, "*fatherless society*". Unfortunately this did not manage to go further, and studies trying to link the social sciences with psychoanalysis remained centred on problems of authority. The development of some states was characterized by a manipulation of the masses. Thus, on the one hand, for example, an enemy designated for them to be overthrown, with the promise of so achieving total happiness, invented "*metaphysical fetishes*" authorizing them to defy the prohibitions of the superego by virtue of clever rationalizations as, for instance, in the Nazi regime or the "socialist" régimes (Burma, Cambodia, and Cuba, among others). On the other hand, established controlling bodies of democratic freedom—like the free press, public opinion, the parliamentary system—were depreciated and ridiculed so that all state control by its citizens would become impossible. Such factors are certainly important, but to explain them it is necessary to recognize that man, far from being the "*noble savage*" that Rousseau desired, is only just the *savage* whose true nature, in which the life instinct rubs shoulders with the death instinct, is hidden beneath a very thin layer of ideals and precarious repression. This central notion of Freudian work is confirmed by other non-psychoanalytic research, which brought to the fore the importance of aggressivity and the fragility of mechanisms controlling it in man, in comparison with what happens in other primates and mammals. Freud emphasizes (1930a) that "the *inclination to aggression is an original, self-subsisting, instinctual disposition in man*, and I return to my view that it *constitutes the greatest impediment to civilization*" (p. 122, italics added).

Because of these tensions and in spite of them, man always hopes for the realization of his most beautiful dream: the advent of the kingdom of God, or, more recently, all that the noblest spirits of our time could imagine: peace, justice, and freedom. Psychoanalysis reveals the aggressivity, envy, and jealousy that disturb the internal worlds of every-day, peaceful, and civilized men. It shows how they just await the cheapest rationalizations in order to be able to discharge their tensions. For social life to be *tolerable* and to prevent the average man from becoming a murderer, there are necessarily conditions enabling a continuous discharge in small doses of competitive feelings creating subliminal tensions. Freud continues (1930a):

> It is clearly not easy for men to give up the satisfaction of this inclination to aggression. They do not feel comfortable without it. The advantage which a comparatively small cultural group offers of *allowing this instinct an outlet in the form of hostility against intruders* is not to be despised. When once the Apostle Paul had posited universal love between men as the foundation of his Christian community, extreme intolerance on the part of Christendom towards those who remained outside it became the inevitable consequence. [p. 114, italics added]

History has tragically confirmed the correctness of his views and has demonstrated that the truest, most realistic, and deepest appreciation of man ought to preside over our social considerations. A thread running through a whole stream of ideas since the Enlightenment (Rousseau's "noble savage", or Marx) tends to suggest that it is exclusively social institutions which degrade men—while in Freud's thinking, it is the instincts (Freud, 1930a):

> The world "civilization" describes the whole sum of the achievements and the regulations which distinguish our lives from those of our animal ancestors and which serve two purposes—namely to protect man against nature and to *adjust their mutual relations.* [p. 89, italics added]

Indeed, the wishes and fears of the *internal world* revealed by psychoanalysis are only partial causes of events. There are *other factors*, whether social or historic, that have a joint effect. For example, terrorism may result when the forces towards change have no other means of expressing themselves. But the question can be asked as to what the intrapsychic forces conducive to terrorism are when other methods would also be available for influencing society. With relation to the analysis of Italian and German terrorist movements, psychoanalysis has tried to understand what personality type will be actualized or utilized by a political current or social datum. The Freudian revolution discovered that the irrational, which had already been grasped by poets—Sophocles, Shakespeare—and later by philosophers—Schopenhauer and Nietzsche—has a *structure* that we can try to *understand*. Criminal acts against what is designated "the enemy" are justified by the projection inherent in fanaticism through the scapegoat mechanism and under the pretext of extirpating evil. Nietzsche observed that *people who are dissatisfied with themselves are always ready to avenge themselves; we others become their victims* (quoted by Feifel, 1965).

It would be ridiculous to consider such books as *Civilization and its Discontents* and *Future of an Illusion* only as ideological texts. In fact, when it comes to passing judgement on communism or the accumulation of capital, these works are amateurish. If, however, an attempt is made, through the Freudian hypothesis, to elucidate the mental functioning of man and his unconscious in a spirit of understanding, it may be said that these works can offer a basis for studying "*homo fanaticus*". As the fruits of clinical experience and a particular vision of man, they can provide for investigating the depths of his unconscious and helping us towards an awareness of the fanatic *in each one of us*. In their book *Freud or Reich?* (1976), Grunberger and Chasseguet-Smirgel write: "We think that psychoanalysis has a quite particular propensity for being converted into ideology."

Historical conditions have contributed to the rebirth of fanatical thinking, which, until the nineteenth century, had been the prerogative of religions. After the Enlightenment and the Encyclopaedist movements, man thought that science is able to pronounce ultimate truths in relation to a rational, and therefore good, form of communal life for human beings. The mere imposition of such truths would suffice to attain utopian happiness. Such ideas have prevailed in dogmatism from the French Revolution to Cambodia, to Stalin and to Ceaucescu. Obviously, other economic factors are also involved—the wish in some countries, for instance, on the part of the intelligentsia, to obtain greater power at the expense of the commercial and working classes, and the alliance of these ideas with the libertarian movements of different on various continents.

Civilization, being more than just ascetic, both organizes pleasure and reduces anxiety in order to offer a life worthy of a man—"worthy of a civilized man"—with the maximum of pleasure *actually* possible. Social problems are not only external problems; society is also the bearer of projections. Problems posed at the level of society as a whole are at the same time individual problems—the internal problems of *the people of whom that society is composed*. They can be grasped through a whole network of concepts referring back to the clinical experience of psychoanalysis. This has been shown by Freud's notion of civilization and his analysis of the processes implied in it.

From a psychoanalytic viewpoint, the *roots of evil* are to be found in the wish for "omnipotence", for the death instinct can also be understood as the regression to a malicious narcissism, which triggers off the destructive aggressivity of man. This omnipotence is linked to the refusal to accept human nature, especially of two points which mark its limits and its animality: sexuality and death. The denial of death and the regression of sexuality correspond to ideals of omnipotence at a level of *individual* neurosis, while fanaticism relates to the death instinct at the cultural level. Anxiety has to do with life; sexuality and joy are associated with anxiety about death. Wilhelm Reich, in *The Mass Psychology of Fascism* (1933), described the dynamics of human misery deriving from man's attempt to be other than he is, to deny his animal nature. This misery is attributed, through profound projective mechanisms, to Jew or Gypsy or, in other systems of reference, to the grasping rapacity of the bourgeoisie. Against them the death instinct dictates the "*final solution*": the death camps or gulags. If he does not accept his imperfection and failings, man must put them outside himself through some paranoid mechanism in order not to be crushed by the weight of his guilt feelings. This contradiction between the life and death instincts is inherent in *fanaticism: destruction* in the name of a search for an *ideal world*.

Indeed, according to Freud (1927c):

> these privations [imposed by civilization] are still operative and still form the kernel of hostility to civilization. . . . Cannibalism alone seems to be universally proscribed and . . . to have been completely surmounted. . . . It is possible that cultural developments lie ahead of us in which the satisfaction of yet other wishes, which are entirely permissible today, will appear just as unacceptable as cannibalism does now. [p. 11]

For Róheim (1943), culture is the system for protecting a child "*who is afraid to stay alone in the dark*". This would seem to be, then, a search for security by a creation—sometimes an illusion—and would derive from the permanent infantile nature of man (Freud, 1927c):

> It is remarkable that, little as men are able to exist in isolation, they should nevertheless feel as a heavy burden the sacrifices which civilization expects of them in order to make a *communal* life possible. . . . Human creations are easily destroyed, and science and technology, which have built them up, can also be used for their annihilation. [p. 6, italics added]

Truly prophetic words.

In *Totem and Taboo* (Freud, 1912–13, pp. 141–143) the instinctual *forces* underlying *civilization* are brought out by the description of a primitive horde in which the omnipotent father appropriates the women. Repression was initially linked to the sexual monopoly and especially the power of the father's domination. The wish to take his attributes and hatred of the frustrations imposed would provoke the first revolution, and the sons would kill the father. But the ambivalence of their feelings for him would prevent their enjoyment of victory. "The dead man became more powerful than ever he was when alive", and the brothers would have to reinstate his power structures and domination. So there would be no chance of real change from the outside, since, when the father is dethroned, the sons, because of their guilt, can only *reproduce the institutions*. This is how Freud (1912–13) describes the repetition compulsion:

> If you want to expel religion from our European civilization, you can only do it by means of another system of doctrines; and such a system would from the outset take over all the psychological characteristics of religion—the same sanctity, rigidity and intolerance, the same prohibition of thought—for its own defence. [p. 51]

One remembers how the French Revolutionaries reinstalled religious forms while ferociously combating religion. Every true revolution would therefore have to coincide with an *internal* change. Without that, there is no escape from the repetition compulsion. Kafka (1974) said that when revolution disappears, what is left is the mass of a new bureaucracy, and that the chains of tortured humanity are forged from wretched papers.

The general aim of the individual's developmental process is the search for happiness according to the pleasure principle. Adaptation to the human community is a prerequisite for the achievement of this happiness. The individual finds himself obliged to integrate two drives: the "egotistic" drive towards happiness, and the "altruistic" one towards association with others in the community. It could be said that the community develops *a collective superego*, which presides over cultural development by *elaborating ideals* and *imposing demands*. Among the latter, those that are concerned with relationships between men are grouped under the term "ethics". The fate of the human species depends on its capacity to succeed, through cultural development, in mastering what disturbs communal life, viz. aggressivity and self-destruction.

Marcuse (1967) sees one solution to the repetition compulsion as watching out for revolutions in the regression towards a "maternal Utopia", whose values would be receptiveness, pacification, gentleness, and sensitivity. His work, like Fromm's, radiates a profound aversion for patriarchal values and confrontation with the "father". Let it be noted that the maternal Utopia presents a regressive character—"intra-uterine"—stripped of aggressivity and was never realized as such until now.

With the Eros/Thanatos dichotomy, Freud tried to define the opposition between the creative and destructive forces on an individual level. The theory of the death instinct was destined to explain the potential presence of evil in man as well as the limits of human life and their inescapability. Since there is always a counterpart of loss in any gain, in some cases, despite the sacrifices that neurosis implies, it might be preferable to the recovery of health. Thus human aims, which are often contradictory, *cannot be realized harmoniously all at the same time*. Ambivalence and tensions are the lot of humanity (cf. also Roazen, 1971).

Ideological science or scientific ideology

An end to ideologies can only delight the friends of Reason—but a reasonable reason, which takes into account the irrationalities of man. It is in this sense that Freud is post-modern—and also in relation to the construction of his theory, which was never globalizing but which proceeded bit by bit by including the contradictions themselves. Freudianism ought not to play a role of replacement ideology, of "*Ersatz*"—Freud was specific about that. He professed that his vision of the world, his "*Weltanschauung*", is quite simply one of science: to *question*. It is in this questioning, in the astonishment and surprises that attend any study of human nature, that psychoanalysis remains topical. It is quite clear that Freudians also have their dogmas and fanaticisms. Among the complex reasons for this, which I have tried to analyse, are: psychoanalysis as a movement among other "isms", Freud's complex relationship with his disciples, his marginal position thirsting for recognition, etc. They all have their rationale. But in other historical circumstances, in our post-modern world, Freudianism may, if it can rid itself of the temptations from the beginning half of the century, play an important role for our civilization, for humanity, in that eternal quest for "*Ignothi seauton*", the Delphic programme: know thyself and what could be termed human nature in all its variety.

For Freud, psychoanalysis discovered the importance of the drives, but it does not assume admiration complacently. The elaboration of demands coming from the drives is helpful to the dynamic victory of reason and mind, the triumph of the Enlightenment. According to Freud (1927c), even if

> man's intellect is powerless in comparison with his instinctual life . . . the voice of the intellect is a soft one, but it does not rest till it has gained a hearing. . . . This is one of the few points on which one may be optimistic about the future of mankind. . . . [p. 53]

Freud's dream was the advent of reason under the influence of the French eighteenth-century philosophers—in particular of Voltaire, the apostle of tolerance. Moreover, it is to Voltaire we owe the expression "fanaticism" in its present meaning.

Freud hoped that man would eventually abandon that archaic mentality which separates the world into Good and Evil. Unfortunately, T. S. Eliot's formulation has never ceased to hold, viz. that "*human kind cannot bear very much reality*".

One must always remember the possibility of illusions gained (Gressot, 1965), temporarily it is true, but which may be succeeded later by knowledge. Illusions are surmountable; they can be structuring at certain times in a child's development, for example, in so far as they do not interfere with the individual's psychological economy.

Experience has shown that an ethic cannot be drawn directly from psychoanalysis. It could only arouse misunderstandings and contradictions. But the transformation of awareness, of the way of looking at man, may lead, as other cultural facts also, to an improvement in our thinking about ourselves and our neighbours. And this may, in turn, lead indirectly to a true revolution, to a change in the quality of man's thinking and talking about himself—*a work of truth* about the difficulties that Freud calls the "*harshnesses of life*". Nietzsche said that man is a "sick animal"; Freud makes manifest the fact that human situations are inescapably conflictual. Man is the only creature who has such a long childhood and who thereby lives in such a long-term dependency. He is linked to his history, characterized by this long infantile dependency and by the great impact of the first experiences he has with the world (Oedipal crisis, castration anxiety) and with people close to him (father, mother, siblings), and with the later re-experiencing of all these elements. Freud calls the repetition compulsion "uncanny". It leads us back untiringly along the same paths, in the image of the Greek dramas in which fate is inevitable, and it goes hand in hand with the conflictual character of man because of the contradictions internalized after his long childhood. The discord between the ego and the sexual and aggressive instincts—conflict between life and death instincts—is part of the inescapable destiny of man. His impaired internal world drives him to the process of working through, and thus he may arrive at a better *integration* and *creativity*. This work can be accomplished in psychoanalytic treatment or in the cultural process; its outcome is *creativity*. At the same time, the psychoanalytic treatments that Freud conducted showed him that we have to live in the shadow of despair.

In the end it is the *lucid* knowledge of the necessary character of conflicts that, if not the last, is at least the first precept of a wisdom that incorporated psychoanalytic teaching. Not only did Freud renew the sources of the tragic in understanding the structure of the irrational, but the "knowing that this is tragic" itself became a possible reconciliation with the inevitable. Freud, the naturalist, the determinist, the scientist, and the inheritor of the Enlightenment, re-found the language of the tragic myths of Oedipus, Narcissus, Eros, Ananké, and Thanatos.

The awareness that psychoanalysis gives, which Freud offers to modern man, is difficult and painful, but in terms of cost it is analogous to the reconciliation of which Aeschylus pronounced the law: "*to pathei mathos*" [understanding through suffering].

The Freudian position ought also to inspire us with courage. Freud was an example of it throughout his life, as witnessed by his restraint in the face of the most tragic events that struck him, among others the loss of several members of his family during the First World War, then the persecution, and finally the long confrontation with death. The stoicism with which he faced death is, in itself, an exemplary attitude.

Envoi

T he reader will have noticed that, except for a clinical overture and an epilogue, this book is essentially in two parts.

The *first* takes up the thread of my preceding work: "Psychoanalytic *controversies* address the issue of psychoanalytic *practice*. . . . Psychoanalytic theory can only be validated by constant reference to experience, to the practice on which it is founded" (Haynal, 1987a, p. 143).

The present volume has set out to study this basis, particularly from the point of view of communication in the psychoanalytic situation—one of its fundamental problems, in fact its practice and the theory deriving from it. Secondarily, it has attempted to clarify the problems of human *communication* and *change*.

Thus it is the epistemological and methodological bases of psychoanalysis that are examined. Each science must retain the *specificity* of its object, its viewpoints, its methodology, its processes, and its practice. At the same time, it must be able to communicate with neighbouring sciences and to participate in the general scientific movement of the time. Otherwise, it is

245

isolated. In its wish to preserve its "purity" it can only be condemned to being deprived of the stimulation presented by such exchanges. To put psychoanalysis in a context of *contemporary scientific exchanges*, therefore, brings obvious advantages, and it is not by chance that some of its representatives have tried to install it in the university—Freud himself had always wished for that. (We should remember how much he strove to become a university professor—the expression of a desire not only for respectability but also to belong to the scientific community.) For some sensitive psychoanalysts, seeking contacts with other sciences *where* they are established and are taught—namely, at the university—is a betrayal of the marginality considered to be the main protection of psychoanalysis and the source of its originality. I do not share this latter opinion but rather submit to the example of Freud, who tried—as his studies in linguistics and ethnology testify—to find his place in the scientific community of his contemporaries.

My propositions will, I hope, *engender* new ideas and encourage a better understanding of the subjects at issue. In addition, I hope that this work will enable a better *dialogue* to take place *between the analysts* on the one hand, and between analysts and *the representatives of other sciences* on the other. This is important, for the future both of psychoanalysis and of the human sciences in general.

* * *

The second part of the book examines the effects on psychoanalysis of cultural *history* and both the scientific and artistic milieux. Is it necessary to recall that psychoanalysis was born out of a fertile exchange with the sciences of the day, in the heart of that melting pot that was Vienna at the end of the nineteenth century? Contemporary historiography has fully emphasized the degree to which it was "in the air of the time". [Similarly with Darwin and Marx. The idea of the evolution of species had already been encountered by Lucretius and by the thinkers of the eighteenth century: Buffon, Diderot, Goethe, Erasmus Darwin (Charles' grandfather), and later, obviously, by Lamarck, Robert Chambers, and Alfred Russell Wallace (Gregory, 1987). A similar case can be made for Marx (Kolakowski, 1978).] Freud's merit was in having built a structured theory, an open system, which is always able to be modified in line with its own "praxis" and with a general revolutionary conception. This no longer needs demonstration. My thesis consists solely in *renewing* the thinking about psychoanalysis in the light of the *praxis* and *ideas* that are drawn both from its own tradition and from contemporary sciences.

The *history of psychoanalysis* is interwoven with the development of practices and theory and the personalities that have been at the source of successive modifications. Previously, scientific psychology had been eminently experimental since Wundt, in the mid-nineteenth century. Pavlov, Watson, and the behaviourists were quite unable to offer a real understanding of man's complex functioning, in so far as they included neither his affectivity, his sexuality, his passions, his dreams, nor his pathology. It should be noted that the "cognitive" movement also errs in brushing aside emotion and experience and thereby remains inadequate for studying human *complexity*. The psychoanalytic model, which is based on a wider ensemble of data and sets out to take account of representation, affect, and experience, offers the advantage of tackling human functioning in all its complexity. Psychoanalysis thus constitutes a rich laboratory for studying such functioning, but the experience it brings has not been sufficiently exploited because of its methodological and epistemological inadequacies. The time has come, therefore, to re-examine these problems in the light of knowledge acquired in the field of human and social sciences (cognitive psychology, ethnology, sociology, and linguistics, to name only the most important) and considerations arising out of the study of information processing, including artificial intelligence.

Psychoanalysis is the result of a sum of experiences of an astonishing richness and an admirable speculative construction about which Freud (1940a [1938]) said modestly: ". . . for the moment we have nothing better at our disposal . . . and for that reason, in spite of its limitations, it should not be despised" (p. 182).

However, this is no longer entirely true today, for there are other models that aim at responding to problems raised in the first place by psychoanalysis.

I have set out to study the most striking concepts in psychoanalytic practice and to evaluate them in a clinical perspective through the observations of a *narrative* that is part of a *process*. My aim has also been to see what such a study can contribute to the understanding of more specific problems, such as *communication*—especially *affective* communication—in general. Lastly, with regard to the latest human sciences and the most advanced information theories, I have tried to open up a fruitful *dialogue* with them to achieve a better understanding of man, his internal world, and his difficulties.

* * *

From the outset and throughout its long history, psychoanalytic thinking has evolved in contact—in dialogue—with its cultural and scientific *circle* and under its influence. In Freud's writings it is abundantly clear; subsequently, references and traces of these influences are implicit rather than explicit, for reasons linked to the sociology of the psychoanalytic movement. Explicit reference to Freud and to the first Freudians is intended to put the permanence of thought back into the continuity of their thinking, while references to non-Freudian influences are often evaded for fear of seeming not to belong in the Freudian scientific tradition, or even of being disloyal to it. However, it is inconceivable that Freudian development should take place without external cultural influences. For example, D. W. Winnicott's work and that of his follower R. Laing is difficult to imagine without the influence of existentialism, that of Jacques Lacan without German phenomenology and (pseudo-) Saussurian linguistics, and that of Bion without Kantian philosophy and the Indian traditions of thought. (The history of these interactions is, moreover, most exciting, as is apparent in the second part of the book.)

Psychoanalysis is undoubtedly being *challenged* in the current scientific context. Taking up such a challenge and contributing to a renewal of psychoanalytic thinking is the aim of this book. If it is referred to in the language of contemporary communication, psychoanalysis is young: created by Freud in a dialogue with the environmental sciences, it has grown in that dialogue, and perhaps it will also reach its maturity at the end of this century in a similar dialogue.

My objective is to relocate psychoanalysis in the present cultural and professional context, to reconsider it from the epistemological and historical angle, and to explore a way of forming or continuing a dialogue with its neighbouring sciences. This is my aim because I share the conviction of one of the founders of modern science, Sir Francis Bacon, who considers that error is less serious than confusion [". . . *citius emergit veritas ex errore quam ex confusione*"] (Bacon, 1620, I, Vol. 4, p. 261). As in histology, I examine the problems encountered through two "sections"—transverse and longitudinal. In the first, synchronic, approach, I look at the problem of psychoanalysis as it is presented today; in the second, diachronic, approach, I relocate this problem in the context of historical development. I hope that my observation and reflection on current practice, on the psychoanalytic *process*, will meet with a wide consensus capable of transcending individual *contents* marked by so many experiences, personal suppositions, and cultural backgrounds.

My programme is to work in the Freudian intellectual tradition by studying it in the light of clinical experiences extending over a hundred or so years, in a new cultural situation, with the emergence of new modes of thought and epistemology.

Through its orientation, this book is a study of problems raised by psychoanalytic theory in an academic perspective. It could not have been achieved without difficulty, bearing in mind the well-known and frequently fierce antagonism that exists between practitioners and researchers. "Practitioners" regard "academics" with a suspicion that has its roots in the safety provided by the rules and habits of thinking. If it is true that through theory we are better able to understand our patients and to progress in their treatment, calling it into question inevitably raises anxiety. But it has to be accepted that progress towards knowledge can only be achieved at a price that has to be paid in terms of intellectual discomfort.

Nowadays, a science can no longer be constructed as a total system but, rather, as a limited *opening* on observable problems with a given methodology. This work hopes to give an account of such openings on the study of human relationships and communication from the point of view of psychoanalytic clinical method, rethought with the conceptual tools available with the approach of the twenty-first century. Some parts of this book are no more than outlines roughly sketched and therefore incomplete. But by that very fact I am following a tradition started by Freud himself as a precursor of *post-modernity*. Any project, any framework, can only leave blanks, empty spaces in the construction of an "edifice" in which later experiences will have to find their place.

This, let me repeat, is how Freud worked, as a pioneer, as a beacon, as "an *elucidator*"—a word that Freud (1895) used to characterize himself:

> One works to be the best of one's power, as an elucidator (where ignorance has given rise to fear), as a teacher, as the representative of a freer or superior view of the world, as a father confessor who gives absolution as it were, by a continuance of his sympathy and respect after the confession has been made. One tries to give the patient human assistance, so far as this is allowed by the capacity of one's own personality and by the amount of sympathy that one can feel for the particular case. [pp. 282–283]

An "elucidator" designates a Socratic person who brings the light of knowledge to his fellow men. He comes from the era of Heine—an era of an extraordinary merging of central European and German-speaking cultures, from which rose Freud, Wittgenstein, Popper, and many others. Freud is first and foremost such a thinker. As such, he stimulated reflection, pursuing his questions, demolishing what he had previously constructed, thus inaugurating the very dynamics of his science. Like Rorty (1980), I, too, prefer "educators" to systematizers—and Freud was, above all, an educator. The "systematizations" of psychoanalysis are not his doing and are not intellectually fertile enterprises. It is within the age-old tradition of questioning, and challenging existent knowledge, that I wish to stand.

REFERENCES

Abraham, K. (1911). Giovanni Segantini: A psychoanalytic study. In: *Clinical Papers and Essays on Psycho-Analysis* (pp. 210–261). London: Hogarth Press, 1955. [London: Karnac Books, 1979.]

Ainsworth, M. D. S., Blehar, M. C., Waters, E., & Wall, S. (1978). *Patterns of Attachment. A Psychological Study of the Strange Situation*. Hillsdale, NJ: Lawrence Erlbaum.

Alexander, F., Eisenstein, S., & Grotjahn, M. (Eds.) (1966). *Psycho-analytic Pioneers*. New York: Basic Books.

Andreas-Salomé, L. (1966). *The Freud Journal of Lou Andreas-Salomé*. London: Hogarth Press.

Anzieu, D. (1959/1975/1988). *Freud's Self-Analysis*. London: Hogarth Press.

Atlan, H. (1979). *Entre le cristal et la fumée. Essai sur l'organisation du vivant*. Paris: Seuil.

Bacon, F. (1620). *The Works of Francis Bacon, Vol. 1*, "new edition" (J. Spelding, R. Leslie, & D. Denon Heath, Eds.). London: 1889.

Balint, M. (1933). Character analysis and new beginning. In: *Primary Love and Psycho-Analytic Technique* (second edition) (pp. 151–164). London: Tavistock Publications, 1965. [Reprinted London: Karnac Books, 1985.]

___ (1934). Dr Sándor Ferenczi as a psycho-analyst. Necrology of 3 October 1933, at the Hungarian Psychoanalytic Society. *Indian Journal of Psychology*, 9: 19–27. [Reprinted in: *Problems of Human Pleasure and Behavior. Classic Essays in Humanistic Psychiatry* (pp. 235–242). New York: Liveright, 1956. Reprinted London: Karnac Books, 1987.]

___ (1965). *Primary Love and Psychoanalytic Technique* (second edition). London: Tavistock Publications.

___ (1968a). *The Basic Fault: Therapeutic Aspects of Regression.* London: Tavistock Publications. [Reprinted London: Karnac Books, 1979.]

___ (1968b). The unobtrusive analyst. In: *The Basic Fault: Therapeutic Aspects of Regression* (pp. 173–181). London: Tavistock Publications. [Reprinted London: Karnac Books, 1979.]

___ (1969). Trauma and object relationship. *International Journal of Psycho-Analysis*, 50 (4): 429–435.

___ (1985 [1969]). Draft Introduction & Notes for a Preface. In: S. Ferenczi, *The Clinical Diary of Sándor Ferenczi* (pp. 219–222). Cambridge, MA: Harvard University Press, 1988.

Balint, M., & Balint, A. (1939). On transference and counter-transference. *International Journal of Psycho-Analysis*, 20 (3–4): 223–230. [Reprinted in: *Primary Love and Psycho-Analytic Technique* (second edition) (pp. 201–208). London: Tavistock Publications, 1965. Reprinted London: Karnac Books, 1985.]

Berger, A. Freiherr von (1896). Surgery of the soul. *Wiener Morgenpresse*, 2 February. [Reprinted in: *Almanach der Psychoanalyse* (pp. 285–289), 1933.]

Binswanger, L. (1956). *Sigmund Freud: Reminiscences of a Friendship.* New York/London: Grune & Stratton.

Bion, W. R. (1957). On arrogance. *International Journal of Psycho-Analysis*, 39: 1958. [Reprinted in: *Second Thoughts, Selected Papers on Psychoanalysis* (pp. 86–92). New York: Jason Aronson, 1967. Reprinted London: Karnac Books, 1984.]

___ (1962). *Learning from Experience.* New York: Basic Books. [Reprinted London: Karnac Books, 1984.]

___ (1965). *Transformations. Change from Learning to Growth.* London: Heinemann. [Reprinted London: Karnac Books, 1984.]

Blanton, S. (1971). *Diary of My Analysis with Freud.* New York: Hawthorn Books.

Bleger, J. (1967). Psycho-analysis of the psycho-analytic frame. *International Journal of Psycho-Analysis*, 48: 511–519.

Bolland, J., & Sandler, J. (1965). *The Hampstead Psychoanalytic Index.* New York: International Universities Press.

Bourguignon, A., Cotet, P., Laplanche, J., & Robert, F. (1989). *Traduire Freud.* Paris: Presses Universitaires de France.

Bowlby, J. (1969). *Attachment and Loss. Vol. 1: Attachment.* London: Hogarth Press.

___ (1973). *Attachment and Loss. Vol. 2: Separation: Anxiety and Anger.* London: Hogarth Press.

___ (1979). Psychoanalysis as art and science. *International Review of Psycho-Analysis, 6:* 3–14.

___ (1980). *Attachment and Loss. Vol. 3: Loss, Sadness and Depression.* London: Hogarth Press.

___ (1981). Psychoanalysis as a natural science. *International Review of Psycho-Analysis, 8:* 243–255.

Brenner, C. (1984). Countertransference as a compromise formation. In: E. Slakter (Ed.), *Countertransference* (pp. 43–51). North Vale, NJ: Jason Aronson, 1987.

Brim, O. G. Jr., & Ryff, C. D. (1970). On the properties of life events. In: P. B. Baltes & O. G. Brim, Jr. (Eds.), *Life-Span Development and Behavior. Vol. 3.* New York: Academic Press.

Brome, V. (1982). *Ernest Jones, Freud's Alter Ego.* London: Caliban Books.

Brown, N. O. (1959). *Life Against Death. The Psychoanalytic Meaning of History.* Middletown, CT: Wesleyan University Press.

Cahn, R. (1983). Le procès du cadre. *Revue Française de Psychanalyse, 47:* 1107–1134.

Calder, K. T. (1980). An analyst's self-analysis. *Journal of the American Psychoanalytic Association, 28:* 5–20.

Canetti, E. (1967). *Torch in My Ear.* Framingham, MA: Deutsch.

Carotenuto, A. (1980). *A Secret Symmetry. Sabina Spielrein Between Freud and Jung.* London: Routledge & Kegan Paul, 1984.

Carus, C. G. (1846). *Psyche, zur Entwicklungsgeschichte der Seele.* Pforzheim: Flammer & Hoffmann.

Chasseguet-Smirgel, J. (1975). *The Ego-Ideal.* London: Free Association Books, 1985.

Dahl, H., Teller, V., Moss, D., & Trujillo, M. (1978). Countertransference examples of the syntactic expression of warded-off contents. *The Psychoanalytic Quarterly, 47* (3): 339–363.

Darwin, C. (1892). In: F. Darwin (Ed.), *The Autobiography of Charles Darwin and Selected Letters.* New York: Dover, 1958.

Dohrenwend, B. S., & Dohrenwend, B. P. (1974). Overview and prospects for research on stressful life events. In: B. S. Dohrenwend & B. P. Dohrenwend (Eds.), *Stressful Life Events: Their Nature and Effects* (pp. 313–331). New York: John Wiley.

Donnet, J. L. (1983). L'enjeu de l'interprétation. *Revue Française de Psychanalyse, 47*: 1135–1150.

Dorey, R., Castoriadis, C., Enriquez, E., et al. (1991). *L'inconscient et la science.* Paris: Dunod.

Dreyfus, H. L. (1979). *What Computers Can't Do: The Limits of Artificial Intelligence.* New York: Harper & Row.

Dubos, R. (1960). *Pasteur and Modern Science.* New York: Anchor.

Dührssen, A. (1972). *Analytische Psychotherapie in Theorie, Praxis und Ergebnissen.* Göttingen: Vandenhoeck und Ruprecht.

Dupont, J. (1985). Introduction. In: S. Ferenczi, *The Clinical Diary of Sándor Ferenczi* (pp. xi–xxvii). Cambridge, MA: Harvard University Press, 1988.

Duruz, N. (1985). *Narcisse en quête de soi.* Brussels: Mardaga.

Eissler, K. R. (1982). Psychologische Aspekte des Briefwechsels zwischen Freud und Jung. *Jahrbuch der Psychoanalyse, Supplement 7.* Stuttgart: Frommann-Holzboog.

Ekman, P. (1973). *Darwin and Facial Expression. A Century of Research in Review.* New York: Academic Press.

___ (1980). L'expression des émotions. *La Recherche, 11* (117): 1408–1415. [Reprinted in: B. Rimé & K. Scherer (Eds.), *Les émotions* (pp. 183–201). Neuchâtel/Paris: Delachaux et Niestlé.]

Ekman, P., & Friesen, W. V. (1969). The repertoire of nonverbal behavior: Categories, origins, usage and coding. *Semiotica, 1*: 49–98.

Ekman, P., Levenson R. W., & Friesen W. V. (1983). Autonomic nervous system activity distinguishes among emotions. *Science, 221*: 1208–1210.

Elias, N. (1969). *The Civilizing Process.* New York: Urizen.

Ellenberger, H. F. (1970). *The Discovery of The Unconscious. The History and Evolution of Dynamic Psychiatry.* New York: Basic Books.

Engel, G. L. (1975). Ten years of self-analysis: Reaction to the death of a twin. *International Journal of Psycho-Analysis, 56*: 23–40.

Erdelyi, M. H. (1988). Issues in the study of unconscious and defense processes. Discussion of Horowitz's comments, with some elaborations. In: M. J. Horowitz (Ed.), *Psychodynamics and Cognition* (pp. 81–94). Chicago, IL: University of Chicago Press.

Erdheim, M. (1982). *Die gesellschaftliche Produktion von Unbewusstheit. Eine Einführung in den ethnopsychoanalytischen Prozess.* Frankfurt am Main: Suhrkamp.

Eysenck, H. J., & Wilson, G. D. (Eds.) (1973). *The Experimental Study of Freudian Theories.* London: Methuen.

Falzeder, E., & Haynal, A. (1989). Heilung durch Liebe? Ein aussergewöhnlicher Dialog in der Geschichte der Psychoanalyse. *Jahrbuch der Psychoanalyse, 24*: 109–127. [English ed.: Healing through love? A unique dialogue in the history of psychoanalysis. *Free Associations, 2* (1): 1–20, 1991.]

Federn, P. (1952). *Ego Psychology and the Psychoses.* New York: Basic Books. [Reprinted London: Karnac Books, 1977.]

Feifel, H. (1965). *The Meaning of Death.* New York: McGraw-Hill.

Ferenczi, S. (1909 [67]). Introjection and transference. In: *First Contributions to Psycho-Analysis* (pp. 35–93). London: Hogarth Press, 1955. [Reprinted London: Karnac Books, 1980.]

___ (1911 [79]). On the organization of the psycho-analytical movement. In: *Final Contributions to the Problems and Methods of Psycho-Analysis.* (pp. 299–307). London: Hogarth Press, 1955. [Reprinted London: Karnac Books, 1980.]

___ (1912). *First Contributions to Psycho-Analysis.* London: Hogarth Press, 1955. [Reprinted London: Karnac Books, 1980.]

___ (1912 [84]). On the definition of introjection. In: *Final Contributions to the Problems and Methods of Psycho-Analysis* (pp. 316–318). London: Hogarth Press, 1955. [Reprinted London: Karnac Books, 1980.]

___ (1912 [92]). Symbolic representation of the pleasure and reality principles in the Oedipus myth. In: *First Contributions to Psycho-Analysis* (pp. 253–269). London: Hogarth Press, 1955. [Reprinted London: Karnac Books, 1980.]

___ (1913 [111]). Stages in the development of the sense of reality. In: *First Contributions to Psycho-Analysis* (pp. 213–239). London: Hogarth Press, 1955. [Reprinted London: Karnac Books, 1980.]

___ (1919 [210]). Technical difficulties in the analysis of a case of hysteria. In: *Further Contributions to the Theory and Technique of Psycho-Analysis* (pp. 189–197). London: Hogarth Press, 1955. [Reprinted London: Karnac Books, 1980.]

___ (1919 [216]). On the technique of psycho-analysis. In: *Further Contributions to the Theory and Technique of Psycho-Analysis* (pp. 177–189). London: Hogarth Press, 1955. [Reprinted London: Karnac Books, 1980.]

___ (1921 [234]). The further development of the active therapy in psycho-analysis. In: *Further Contributions to the Theory and Technique of Psycho-Analysis* (pp. 198–217). London: Hogarth Press, 1955. [Reprinted London: Karnac Books, 1980.]

___ (1924 [268]). Thalassa: A theory of genitality. *The Psychoanalytic Quarterly*, 2: 361–403, 1933; 3: 1–29 and 200–222, 1934. [Reprinted London: Karnac Books, 1989.]

___ (1928 [282]). The problem of the termination of the analysis. *Final Contributions to the Problems and Methods of Psycho-Analysis* (pp. 77–86). London: Hogarth Press, 1955. [Reprinted London: Karnac Books, 1980.]

___ (1928 [283]). The elasticity of psychoanalytical technique. *Final Contributions to the Problems and Methods of Psycho-Analysis* (pp. 87–101). London: Hogarth Press, 1955. [Reprinted London: Karnac Books, 1980.]

___ (1928 [306]). Über den Lehrgang des Psychoanalytikers. In: *Bausteine zur Psychoanalyse, Vol 3: Arbeiten aus den Jahren 1908–1933* (pp. 422–431). Bern: Huber, 1964. (Not included in English editions of Ferenczi's works.)

___ (1929 [287]). The unwelcome child and his death-instinct. *Final Contributions to the Problems and Methods of Psycho-Analysis* (pp. 102–107). London: Hogarth Press. [Reprinted London: Karnac Books, 1980.]

___ (1930 [291]). The principle of relaxation and neo-catharsis. *Final Contributions to the Problems and Methods of Psycho-Analysis* (pp. 108–125). London: Hogarth Press, 1955. [Reprinted London: Karnac Books, 1980.]

___ (1931 [292]). Child analysis in the analysis of adults. *Final Contributions to the Problems and Methods of Psycho-Analysis* (pp. 126–142). London: Hogarth Press, 1955. [Reprinted London: Karnac Books, 1980.]

___ (1932 [308, in 309]). Notes and fragments. *Final Contributions to the Problems and Methods of Psycho-Analysis* (pp. 216–279). London: Hogarth Press, 1955. [Reprinted London: Karnac Books, 1980.]

___ (1933 [294]). Confusion of tongues between adults and the child. *Final Contributions to the Problems and Methods of Psycho-Analysis* (pp. 156–167). London: Hogarth Press, 1955. [Reprinted London: Karnac Books, 1980.]

___ (1934 [296]). On the revision of the interpretation of dreams. In: Notes and Fragments. *Final Contributions to the Problems and Methods of Psycho-Analysis* (pp. 238–243). London: Hogarth Press, 1955. [Reprinted London: Karnac Books, 1980.]

___ (1985 [1932]). *The Clinical Diary of Sándor Ferenczi* (J. Dupont, Ed.). Cambridge, MA: Harvard University Press.

Ferenczi, S., & Rank, O. (1924 [264]). *The Development of Psychoanalysis*. New York: Nervous and Mental Disease Publishing, Monograph Series 125. [Reprinted London: Karnac Books, 1986.]

Ferenczi, S., & Groddeck, G. (1982). *Briefwechsel 1921–1933*. Frankfurt am Main: Fischer.

Fernald, A. (1989). Intonation and communicative intent in mothers' speech to infants: Is the melody the message? *Child Development, 60* (6): 1497–1510.

Fischer, D. (1977). *Les analysés parlent*. Paris: Stock.

Fischer, R., & Fischer, E. (1977). Psychoanalyse in Russland. In: D. Eicke (Ed.), *Die Psychologie des 20. Jahrhunderts, Vol. 3* (pp. 122–124). Zurich: Kindler.

Fisher, S., & Greenberg, R. P. (1977). *The Scientific Credibility of Freud's Theories and Therapy*. New York: Basic Books.

Flournoy, O. (1968). Du symptôme au discours. *Revue Française de Psychanalyse, 32* (5–6): 807–889.

___ (1989). La science, un danger pour la psychanalyse. In: C. Le Guen, O. Flournoy, I. Stengers, & J. Guillaumin, *La psychanalyse, une science? VIIèmes Rencontres psychanalytiques d'Aix-en-Provence, 1988* (pp. 45–157). Paris: Les Belles Lettres.

Fodor, J. (1983). *The Modularity of Mind. An Essay on Faculty Psychology*. Cambridge, MA: Massachusetts Institute of Technology.

Foucault, M. (1967). *Madness and Civilization: A History of Insanity in the Age of Reason*. London: Tavistock Publications.

Freud, A. (1969). Difficulties in the path of psychoanalysis. In: *The Writings of Anna Freud, Vol. 7*. New York: International Universities Press, 1971.

Freud, S. (1963 [1884–1887]). *The Cocaine Papers* (R. Byck, Ed.). New York: Stonehill, 1975.

___ (1884e). On coca. In: *The Cocaine Papers* (R. Byck, Ed.). New York: Stonehill, 1975.

___ (1884g). Cocaine (unsigned). *Medical News, Vol. 45* (p. 502). Philadelphia. [Reprinted in: A. Grinstein, Freud's First Publications in America. *Journal of the American Psychoanalytic Association, 19* (1971) (2): 255–256.]

___ (1885a). Contribution to the knowledge of the effect of cocaine. In: *The Cocaine Papers* (R. Byck, Ed.). New York: Stonehill, 1975.

___ (1885b). On the general effect of cocaine. In: *The Cocaine Papers* (R. Byck, Ed.). New York: Stonehill, 1975.

___ (1885e). Opinion on Parke's cocaine. In: *The Cocaine Papers* (R. Byck, Ed.). New York: Stonehill, 1975.

___ (1887d). Craving for and fear of cocaine. In: *The Cocaine Papers* (R. Byck, Ed.). New York: Stonehill, 1975.

___ (1888b). "Hysteria" and "hystero-epilepsy". In: *Standard Edition, 1* (pp. 37–57 and 58–59).

___ (1891b). *On Aphasia, a Critical Study.* London: Imago Publishing, 1953; New York: International Universities Press, 1953.

___ (1895d) (with J. Breuer). *Studies on Hysteria. Standard Edition, 2.*

___ (1896a). Heredity and the aetiology of the neuroses. In: *Standard Edition, 3* (pp. 141–156).

___ (1896b). Further remarks on the neuro-psychoses of defence. In: *Standard Edition, 3* (pp. 157–185).

___ (1896c). The aetiology of hysteria. In: *Standard Edition, 3* (pp. 187–221).

___ (1898a). Sexuality in the aetiology of the neuroses. In: *Standard Edition, 3* (pp. 259–285).

___ (1899a). Screen memories. In: *Standard Edition, 3* (pp. 299–322).

___ (1900a). *The Interpretation of Dreams. Standard Edition, 4–5.*

___ (1901a). *On Dreams.* In: *Standard Edition, 5* (pp. 629–686).

___ (1904a). Freud's psycho-analytic procedure. In: *Standard Edition, 7* (pp. 247–254).

___ (1905a). On psychotherapy. In: *Standard Edition, 7* (pp. 255–268).

___ (1905c). *Jokes and their Relation to the Unconscious. Standard Edition, 8.*

___ (1905d). *Three Essays on the Theory of Sexuality.* In: *Standard Edition, 7* (pp. 123–243).

___ (1905e [1901]). Fragment of an analysis of a case of hysteria. In: *Standard Edition, 7* (pp. 1–122).

___ (1906c). Psycho-analysis and the establishment of the facts in legal proceedings. In: *Standard Edition, 9* (pp. 97–114).

___ (1907b). Obsessive actions and religious practices. In: *Standard Edition, 9* (pp. 115–127).

___ (1908c). On the sexual theories of children. In: *Standard Edition, 9* (pp. 205–226).

___ (1908d). "Civilized" sexual morality and modern nervous illness. In: *Standard Edition, 9* (pp. 177–204).

___ (1908e). Creative writers and day-dreaming. In: *Standard Edition, 9* (pp. 141–153).

___ (1909b). Analysis of a phobia in a five-year-old boy. In: *Standard Edition, 10* (pp. 1–147).

___ (1909c). Family romances. In: *Standard Edition, 9* (pp. 235–241).

___ (1909d). Notes upon a case of obsessional neurosis. In: *Standard Edition, 10* (pp. 151–249).

___ (1910c). *Leonardo da Vinci and a Memory of his Childhood.* In: *Standard Edition, 11* (pp. 57–137).

___ (1910d). The future prospects of psycho-analytic therapy. In: *Standard Edition, 11* (pp. 141–151).

___ (1910e). The antithetical meaning of primal words. In: *Standard Edition, 11* (pp. 153–161).

___ (1910i). The psycho-analytic view of psychogenic disturbance of vision. In: *Standard Edition, 11* (pp. 209–218).

___ (1910k). "Wild" psycho-analysis. In: *Standard Edition, 11* (pp. 219–227).

___ (1911b). Formulations on the two principles of mental functioning. In: *Standard Edition, 12* (pp. 213–226).

___ (1912b). The dynamics of transference. In: *Standard Edition, 12* (pp. 97–108).

___ (1912e). Recommendations to physicians practising psycho-analysis. In: *Standard Edition, 12* (pp. 109–120).

___ (1912–13). *Totem and Taboo.* In: *Standard Edition, 13* (pp. 1–161).

___ (1913c). On beginning the treatment. In: *Standard Edition, 12* (pp. 121–144).

___ (1913j). The claims of psycho-analysis to scientific interest. In: *Standard Edition, 13* (pp. 163–190).

___ (1914c). On narcissism: An introduction. In: *Standard Edition, 14* (pp. 67–102).

___ (1914d). On the history of the psycho-analytic movement. In: *Standard Edition, 14* (pp. 1–66).

___ (1914g). Remembering, repeating and working-through. In: *Standard Edition, 12* (pp. 145–156).

___ (1915a). Observations on transference-love. In: *Standard Edition, 12* (pp. 157–171).

___ (1915b). Thoughts for the times on war and death. In: *Standard Edition, 14* (pp. 273–300).

___ (1915c). Instincts and their vicissitudes. In: *Standard Edition, 14* (pp. 109–140).

___ (1915e). The unconscious. In: *Standard Edition, 14* (pp. 159–215).

___ (1916a). On transience. In: *Standard Edition, 14* (pp. 303–307).

___ (1916–17). *Introductory Lectures on Psycho-Analysis. Standard Edition, 15 & 16.*

___ (1917e [1915]). Mourning and melancholia. In: *Standard Edition, 14* (pp. 237–258).

___ (1918b). From the history of an infantile neurosis. In: *Standard Edition, 17* (pp. 1–122).

___ (1919a [1918]). Lines of advance in psycho-analytic therapy. In: *Standard Edition, 17* (pp. 159–168).

___ (1920g). *Beyond the Pleasure Principle.* In: *Standard Edition, 18* (pp. 7–64).

___ (1922b). Some neurotic mechanisms in jealousy, paranoia and homosexuality. In: *Standard Edition, 18* (pp. 221–232).

___ (1923a [1922]). Two enclycopaedia articles. In: *Standard Edition, 17* (pp. 235–259).

___ (1923b) *The Ego and the Id. Standard Edition, 19.*

___ (1923e). The infantile genital organization. In: *Standard Edition, 19* (pp. 139–145).

___ (1925a). A note upon the "mystic writing-pad". In: *Standard Edition, 19* (pp. 225–232).

___ (1925d [1924]). *An Autobiographical Study.* In: *Standard Edition, 20* (pp. 1–70).

___ (1925h). Negation. In: *Standard Edition, 19* (pp. 233–239).

___ (1926d [1925]). *Inhibitions, Symptoms and Anxiety.* In: *Standard Edition, 20* (pp. 75–174).

___ (1926e). *The Question of Lay Analysis.* In: *Standard Edition, 20* (pp. 177–250).

___ (1927a). Postscript to *The Question of Lay Analysis.* In: *Standard Edition, 20* (pp. 251–258).

___ (1927c). *The Future of an Illusion.* In: *Standard Edition, 21* (pp. 1–56).

___ (1930a). *Civilization and its Discontents.* In: *Standard Edition, 21* (pp. 57–145).

___ (1933a). *New Introductory Lectures on Psycho-Analysis.* In: *Standard Edition*, 22 (pp. 5–182).

___ (1933b). *Why War?* (Einstein and Freud). In: *Standard Edition*, 22 (pp. 195–215).

___ (1935b). The subtleties of a faulty action. In: *Standard Edition*, 22 (pp. 231–235).

___ (1937c). Analysis terminable and interminable. In: *Standard Edition*, 23 (pp. 209–254).

___ (1937d). Constructions in analysis. In: *Standard Edition*, 23 (pp. 255–269).

___ (1939a [1937–39]. *Moses and Monotheism.* In: *Standard Edition*, 23 (pp. 1–137).

___ (1940a [1938]). *An Outline of Psycho-Analysis.* In: *Standard Edition*, 23 (pp. 139–207).

___ (1940e [1938]). Splitting of the ego in the process of defence. In: *Standard Edition*, 23 (pp. 271–278).

___ (1941d [1921]). Psycho-analysis and telepathy. In: *Standard Edition*, 18 (pp. 175–194).

___ (1950 [1887–1902]). *The Origins of Psycho-Analysis.* Letters to Wilhelm Fliess, Drafts and Notes 1887–1902. London: Imago, 1954; New York: Basic Books, 1954. Parts in: *Standard Edition*, 1 (pp. 173–397) (including "A Project for a Scientific Psychology", 1895).

___ (1950a [1895]). A project for a scientific psychology. In: *Standard Edition*, 1 (pp. 283–397).

___ (1950a [1897]). Draft M. Notes II (May 25, 1897); Draft N. Notes II (May 31, 1897). In: *Standard Edition*, 1 (pp. 250–253, 254–257).

___ (1955). *The Standard Edition of the Complete Psychological Works of Sigmund Freud.* Translated from the German under the General Editorship of James Strachey. London: Hogarth Press.

___ (1960a). *Letters, 1873–1939.* New York: Basic Books, 1960; London: Hogarth Press, 1961.

___ (1961a). Jung, C. G.: *The Freud/Jung Letters. The Correspondence Between Sigmund Freud and Carl Gustav Jung* (W. McGuire, Ed.). Princeton, NJ: Princeton University Press, 1974.

___ (1963a). *Psycho-Analysis and Faith. The Letters of Sigmund Freud and Oskar Pfister, 1909–1939.* London: Hogarth Press, 1963; New York: Basic Books, 1963.

___ (1965a), Abraham, K., *A Psycho-Analytic Dialogue: The Letters of Sigmund Freud and Karl Abraham (1907–1926)* (H. Abraham & E. Freud, Eds.). New York: Basic Books.

___ (1974). *L'homme aux rats. Journal d'une analyse.* Paris: Presses Universitaires de France.

Freud, S., & Ferenczi, F. (forthcoming). *Correspondence.* (For the original letters, Vienna: Böhlau; for English edition, New York: Harvard University Press.)

Gadamer, H. G. (1975). *Truth and Method.* New York: Seabury Press [orig. *Wahrheit und Methode* (third edition), 1972.]

Gardner, H. (1987). *The Mind's New Science. A History of the Cognitive Revolution.* New York: Basic Books.

Gellner, E. (1988). Psychoanalysis as a social institution. In: E. Timms & N. Segal (Eds.), *Freud in Exile* (pp. 223–229). New Haven/London: Yale University Press.

Genesereth, M. R., & Nilsson, N. J. (1987). *Logical Foundations of Artificial Intelligence.* Los Altos, CA.: M. Kaufmann-Freeman.

Girard, C. (1984). La part transmise. *Revue Française de Psychanalyse, 48* (1): 219–238.

Glover, E. (1955). *The Technique of Psycho-Analysis.* New York: International Universities Press, 1955; London: Baillière, Tindall, 1955.

Goldbeck, Th., Tolkmitt, F., & Scherer, K. R. (1988). Experimental studies on vocal communication. In: *Facets of Emotion. Recent Research* (pp. 119–137). Hillsdale, NJ: Lawrence Erlbaum.

Goodall, J. (1986). *The Chimpanzees of Gomb. Patterns of Behavior.* Cambridge, MA: Belknap.

Gordon, S. L. (1989). Socialization of children's emotions: Emotional culture, competence and exposure. In: C. Saarni & P. L. Harris, (Eds.), *Children's Understanding of Emotions.* New York: Cambridge University Press.

Granoff, W. (1961). Ferenczi: faux problème ou vrai malentendu. *La Psychanalyse, 6*: 255–282.

Green, A. (1967). Le narcissisme primaire: structure ou état? *L'Inconscient, 1* (1): 127–156; *1* (2): 89–116.

___ (1973). *Le discours vivant. La conception psychanalytique de l'affect.* Paris: Presses Universitaires de France.

___ (1977). Transcription d'origine inconnue. *Nouvelle Revue de Psychanalyse, 16*: 27–63.

___ (1979). L'angoisse et le narcissisme. *Revue Française de Psychanalyse, 43* (1): 45–88.

___ (1983). *Narcissisme de vie, narcissisme de mort.* Paris: Minuit.

Gregory, R. L. (1987). *The Oxford Companion to the Mind.* Oxford: Oxford University Press.

Gressot, M. (1965). Les illusions gagnées. Réflexions sur la dualité fonctionnelle, structurante et défensive, des processus de rationalisation. *L'Evolution Psychiatrique, 30* (4): 577–608.

Groddeck, G. (1974). *The Meaning of Illness. Selected Psycho-Analytic Writings, including His Correspondence with Sigmund Freud.* London: Hogarth, 1977. [Reprinted London: Karnac Books, 1987.]

Grosskurth, Ph. (1986). *Melanie Klein. Her World and Her Work.* New York: Alfred A. Knopf. [Reprinted London: Karnac Books, 1987.]

Grünbaum, H. (1984). *The Foundations of Psychoanalysis: A Philosophical Critique.* Berkeley/Los Angeles/London: University of California Press.

Grunberger, B. (1957). Situation analytique et processus de guérison. *Revue Française de Psychanalyse, 21*: 375–458.

___ (1971). *Narcissism. Psychoanalytic Essays.* New York: International Universities Press, 1979.

___ (1974). De la technique active à la confusion de langues. *Revue Française de Psychanalyse, 38*: 521–546.

Grunberger, B., & Chasseguet-Smirgel, J. (1976). *Freud or Reich? Psychoanalysis and Illusion.* London: Free Association Books, 1986.

Gusdorf, G. (1984). *L'homme romantique.* Paris: Payot.

___ (1988). *Les origines de l'herméneutique.* Paris: Payot.

Guttman, S. A., Jones, R. L., & Parrish, S. M. (1980). *The Concordance to the Standard Edition of the Complete Psychological Works of Sigmund Freud.* Boston, MA: Hall.

Habermas, J. (1985). *Philosophical Discourse of Modernity.* Oxford: Polity Press, 1988.

Haeckel, E. (1899). *Les énigmes de l'univers.* Paris: Schleicher, 1901.

Harnik, J. (1923). Schicksale des Narzismus bei Mann und Weib. *Internationale Zeitschrift für Psychoanalyse, 9*: 278–296.

Hartmann, E. von (1869). *Philosophie des Unbewussten.* Berlin: Dunker.

Hartmann, H. (1947). On rational and irrational action. In: *Essays on Ego-Psychology. Selected Problems in Psychoanalytic Technique* (second edition) (pp. 37–68). New York: International Universities Press, 1965.

Haynal, A. (1976). *Depression and Creativity*. New York: International Universities Press, 1985.

___ (1978). Some reflexions on depressive affect. *International Journal of Psycho-Analysis, 59*: 165–171.

___ (1978). Psychoanalytic discourse on orphans and deprivation. In: M. Eisenstadt, P. Rentchnick, A. Haynal, & P. de Senarclens, *Parental Loss and Achievement* (pp. 135–190). Madison, CT: International Universities Press, 1989.

___ (1983a). The psychoanalyst and the psychoanalytic encounter. *Bulletin of the European Psychoanalytic Federation, 20–21*: 19–31.

___ (1983b). La psychanalyse dans le creuset de l'Europe centrale. In: Présence de l'Europe centrale. Le colloque de Duino. *Cadmos, Genève, 6*: 75–86.

___ (1987a). *The Technique at Issue. Controversies in Psychoanalysis from Freud and Ferenczi to Michael Balint*. London: Karnac Books, 1988. [U.S. Edition: *Controversies in Psychoanalytic Method. From Freud and Ferenczi to Michael Balint*. New York: New York University Press, 1989.]

___ (1987b). *Dépression et créativité*. Lyon: Césura.

___ (1989a). The concept of trauma and its present meaning. *International Review of Psycho-Analysis, 16*: 315–321.

___ (1989b). Wie das Leben, so ist auch die Analyse voller Paradoxe. *Zeitschrift für Psychoanalytische Theorie und Praxis, 4* (1): 79–93.

Haynal, A., Molnar, M., & de Puymège, G. (1980). *Fanaticism. A Historical and Psychoanalytical Study*. New York: Schocken, 1983.

Heimann, P. (1950). On counter-transference. *International Journal of Psycho-Analysis, 31*: 81–84.

Heine, H. *Complete Poems*. Oxford: Oxford University Press, 1984.

Herbart, J. F. (1816). *Lehrbuch zur Psychologie*. Königsberg/Leipzig: A. W. Unzer.

___ (1824). Psychologie als Wissenschaft neugegründet auf Erfahrung. Metaphysik und Mathematik. In: *Sämtliche Werke, Vols. 5 & 6*. Leipzig: Voss, 1850.

Hofstädter, D. R. (1987). Cognition, subcognition. *Le Débat, 47* (November–December).

Holmes, Th. H., & Rahe, R. H. (1967). The Social Readjustment Rating Scale. *Journal of Psychosomatic Research, 11*: 213–218.

Horney, K. (1942). *Self-Analysis*. New York: W. W. Norton.

Horowitz, M. J. (1988a). *Introduction to Psychodynamics. A New Synthesis.* New York: Basic Books.

___ (1988b). *Psychodynamics and Cognition.* Chicago, IL: University of Chicago Press.

Jaccard, R. (Ed.) (1982). *Histoire de la psychanalyse.* Paris: Hachette.

Jacoby, R. (1983). *The Repression of Psychoanalysis.* New York: Basic Books.

Janik, A., & Toulmin, S. (1973). *Wittgenstein's Vienna.* New York: Simon & Schuster.

Johnston, W. M. (1980). *Vienna . . . Vienna la Capitale della Nostalgia che ha Inventato il Nostro Presente (1815–1914).* Milano: Mondadori.

Jones, E. (1953). *Sigmund Freud: Life and Work. Vol. 1: The Young Freud, 1856–1900.* London: Hogarth Press.

___ (1955). *Sigmund Freud: Life and Work. Vol. 2: Years of Maturity, 1901–1919.* London: Hogarth Press.

___ (1957). *Sigmund Freud: Life and Work. Vol. 3: The Last Phase, 1919–1939.* London: Hogarth Press.

___ (1959). *Free Associations. Memories of a Psycho-Analyst.* New York: Basic Books.

József, A. (1936). *Poèmes choisis adaptés du hongrois* (introduced by Guillevic). Budapest: Corvinat, 1961.

___ (1956). *Összes versei.* Budapest: Szépirodalmi Könyvkiadó.

Jung, C. G. (1962). *Memories, Dreams, Reflections.* London: Collins, 1963.

Kafka, F. (1974). *Letters to Ottla and the Family.* New York: Schocken, 1982.

Kaufmann, W. (1980). *Discovering the Mind, Vol 3.* New York: McGraw-Hill.

Kernberg, O. F., Burstein, E. D., Coyne, L., Appelbaum, A., Horwitz, L., & Vorth, H. (1972). Psychotherapy and psychoanalysis. Final Report of the Menninger Foundation's Psychotherapy Research Project. *Bulletin of the Menninger Clinic, 36*: 1–275.

Khan, M. M. R. (1963). Ego-distortion, cumulative trauma and the role of reconstruction in the analytic situation. *International Journal of Psycho-Analysis, 45*, 1964. [Reprinted in: *The Privacy of the Self* (pp. 59–68). London: Hogarth Press.]

Kiell, N. (1988). *Freud Without Hindsight.* Madison, CT: International Universities Press.

Klein, M. (1940). Mourning and its relation to manic-depressive states. *International Journal of Psycho-Analysis*, *21*: 125–153. Also in: *Love, Guilt and Reparation and Other Works, 1921–1925*. London: Hogarth Press, 1975. [Reprinted London: Karnac Books, 1992.]

Kline, P. (1972). *Fact and Phantasy in Freudian Theory*. London: Methuen.

Koestler, A. (1952). *An Arrow in the Blue*. New York: Macmillan.

___ (1954). *The Invisible Writing*. Boston, MA: Beacon Press.

___ (1964). *The Act of Creation*. London: Hutchinson.

Kohon, G. (1986). Countertransference: An independent view. In: G. Kohon (Ed.), *The British School of Psychoanalysis. The Independent Tradition* (pp. 51–80). London: Free Association Books.

Kohut, H. (1971). *The Analysis of the Self*. New York: International Universities Press.

Kolakowski, L. (1978). *Main Currents of Marxism*. Oxford: Clarendon Press.

Kramer, M. K. (1959). On the continuation of the analytic process after psychoanalysis (a self-observation). *International Journal of Psycho-Analysis*, *40*: 17–25.

Kraus, K. (1977). *Ausgewählte Werke*. Munich: Langen Mülzer.

Kuhn, Th. S. (1962). *The Structure of Scientific Revolutions*. Chicago, IL: Chicago University Press.

Lacan, J. (1966). *Ecrits. A Selection*. London: Tavistock Publications, 1977.

___ (1973). Du sujet supposé savoir, de la dyade première et du bien. In: *Le Séminaire, Vol. 11* (pp. 209–220). Paris: Seuil.

Lagache, D. (1938). Deuil maniaque. In: *Oeuvres complètes, Vol. 1 1932–1946* (pp. 225–242). Paris: Presses Universitaires de France, 1977.

Langs, R. (1979). The interactional dimension of countertransference. In: L. Epstein & A. H. Feiner (Eds.), *Countertransference* (pp. 71–103). New York: Jason Aronson.

Laplanche, J. (1970). *Life and Death in Psychoanalysis*. Baltimore, MD: Johns Hopkins University Press, 1976.

___ (1982). Carrefour-débat: Biologie et psychanalyse. *Psychanalyse à l'Université*, *7* (28) (September): 532–560.

___ (1987). *New Foundations for Psychoanalysis*. Oxford: Blackwell, 1989.

Laplanche, J., & Pontalis, J. B. (1967): *The Language of Psychoanalysis*. London: Hogarth Press, 1973. [Reprinted London: Karnac Books, 1988.]

Lasch, Ch. (1978). *The Culture of Narcissism. American Life in an Age of Diminishing Expectations.* New York: W. W. Norton.

Lebovici, S. (1960). La relation objectale chez l'enfant. *La Psychiatrie de l'Enfant, 3*: 147–226.

___ (1983). *Le nourrisson, la mère et le psychanalyste. Les interactions précoces.* Paris: Centurion.

Le Guen, C. (1972). Quand le père a peur: ou comment Freud, résistant à son fantasme, a institué les Sociétés psychanalytiques. *Etudes Freudiennes, 5–6.*

Le Guen, C., Flournoy, O., Stengers, I., & Guillaumin, J. (1988). *La psychanalyse, une science? VIIèmes Rencontres psychanalytiques d'Aix-en-Provence.* Paris: Les Belles Lettres.

Leupold-Löwenthal, H. (1980). Freud und das Judentum. *Sigmund Freud House Bulletin, 4* (1): 32–41.

Lévi-Strauss, C. (1955). *Tristes tropiques.* London: Cape, 1973.

___ (1958). *Structural Anthropologie.* New York: Basic Books, 1963/1976.

Lewin, B. (1958). *Dreams and the Uses of Regression.* New York: International Universities Press.

Lieberman, J. (1985). *Acts of Will. The Life and Work of Otto Rank.* New York: The Free Press.

Lindenmann, E. (1944). The symptomatology and management of acute grief. *American Journal of Psychiatry, 101*: 141–148.

Little, M. (1951). Counter-transference and the patient's response to it. *International Journal of Psycho-Analysis, 23*: 32–40.

Loewenstein, R. (1958). Remarks on some variations in psychoanalytic technique. *International Journal of Psycho-Analysis, 39*: 202–210.

Loewenstein, R., Newman, L. M., Schur, M., & Solnit, A. J. (1966). *Psychoanalysis: A General Psychology. Essays in Honor of Heinz Hartmann.* New York: International Universities Press.

Lothar, E. (1960). *Das Wunder des Ueberlebens. Erinnerungen und Erlebnisse.* Hamburg/Vienna: Paul Zsolnay.

Luborsky, L., Crits-Christoph, P., Mintz, J., & Auerbach, A. (1988). *Who Will Benefit from Psychotherapy? Predicting Therapeutic Outcomes.* New York: Basic Books.

Luquet, P. (1962). Les identifications précoces dans la structuration et la restructuration du Moi. *Revue Française de Psychanalyse, 26*: 117–301.

Luria, A. (1925). Psychoanalytische Bewegung. Die Psychoanalyse in Russland. *Internationale Zeitschrift für Psychoanalyse, 11*: 395–398.

Lyotard, J. F. (1979). *The Postmodern Condition. A Report on Knowledge.* Minneapolis: University of Minnesota Press, 1984.

Malan, D. (1965). *A Study of Brief Psychotherapy.* New York: Plenum, 1963/1975.

___ (1976). *The Frontier of Brief Psychotherapy.* New York: Plenum.

Mannoni, M. (1979). *La théorie comme fiction. Freud, Groddeck, Winnicott, Lacan.* Paris: Seuil.

Marcuse, H. (1967). *Five Lectures; Psychoanalysis, Politics and Utopia.* Boston: Beacon Press, 1970.

Marshall, J. C. (1984). Multiple perspectives on modularity. *Cognition, 7:* 209–242.

Marty, P., & M'Uzan, M. de (1963). La pensée opératoire. *Revue Française de Psychanalyse, 27* (supplement): 345–356.

Marziali, E. A. (1984). Prediction of outcome of brief psychotherapy from therapist interpretative interventions. *Archives of General Psychiatry, 41:* 301–304.

Masson, J. M. M. (1984). *The Assault on Truth. Freud's Suppression of the Seduction Theory.* New York: Farrar, Straus & Giroux.

___ (1985). *The Complete Letters of Sigmund Freud to Wilhelm Fliess 1887–1904.* Cambridge, MA/London: The Belknap Press of Harvard University Press.

___ (1988). *Against Therapy. Emotional Tyranny and the Myth of Psychological Healing.* New York: Atheneum.

McCarthy, R. A., & Warrington, E. K. (1988). Evidence for modality-specific meaning systems in the brain. *Nature, 334:* 428–430.

McGuire, W. (1983). Foreword. In: A. Carotenuto, *A Secret Symmetry.* New York: Random House, 1984.

Medawar, P. B. (1967). *The Art of the Soluble.* London: Methuen.

Minsky, M. (1975). A framework for representing knowledge. In: P. H. Winston (Ed.), *The Psychology of Computer Vision.* New York: McGraw-Hill.

___ (1985). *The Society of Mind.* New York: Simon & Schuster.

Moser, U., von Zeppelin, I., & Schneider, W. (1969). Computer simulation of a model of neurotic defence processes. *International Journal of Psycho-Analysis, 50:* 53–64.

Musil, R. (1952). *The Man without Qualities: A Critical Study.* Cambridge, MA: Cambridge University Press.

Myerson, P. G. (1960). Awareness and stress. Post-psychoanalytic utilization of insight. *International Journal of Psycho-Analysis, 41:* 147–156.

Natterson, J. M. (1966). Theodor Reik b. 1888. Masochism in modern man. In: F. Alexander, S. Eisenstein, & M. Grotjahn (Eds.), *Psychoanalytic Pioneers* (pp. 249–264). New York: Basic Books.

Nemiah, J. C., & Sifneos, P. E. (1970). Psychosomatic illness: A problem in communication. *Psychotherapy and Psychosomatics, 18*: 154–160.

Neyraut, M. (1974). *Le transfert.* Paris: Presses Universitaires de France.

Nielsen, E. (1982). *Focus on Vienna 1900. Change and Continuity in Literature, Music, Art, and Intellectual History.* Munich: Houston German Studies, W. Fink.

Nietzsche, F. (1886). *Beyond Good and Evil.* Harmondsworth, Middlesex: Penguin Books, 1990.

___ (1966). *Werke in drei Bänden* (Karl Schlechta, Ed.). Munich: Carl Hanser.

Nunberg, H., & Federn, E. (Eds.) (1962). *Minutes of the Vienna Psychoanalytic Society. Vol. 1: 1906–1908.* New York: International Universities Press.

___ (1967). *Minutes of the Vienna Psychoanalytic Society. Vol. 2: 1908–1910.* New York: International Universities Press.

Paskauskas, R. A. (1988). Freud's break with Jung: The crucial role of Ernest Jones. *Free Associations, 11*: 7–34.

Paul, J. (1804). *Poétique, ou introduction à l'esthétique.* Paris: Durand, 1862. [Reprinted in: *Cours préparatoire d'esthétique.* Lausanne: L'Age d'Homme, 1979.]

Pennybacker, J. W. (1988). Disclosure of traumas and immune function: Health implications for psychotherapy. *Journal of Consulting Psychology, 56* (2): 239–245.

Piaget, J. (1971). The affective unconscious and the cognitive unconscious. *Journal of the American Psychoanalytic Association, 21*: 249–261.

___ (1973). Remarques sur l'éducation mathématique. *Math Ecole, 12* (58): 1–7.

Pickworth Farrow, E. (1926). A method of self-analysis. *British Journal of Medical Psychology, 5*: 106–118.

Plutchik, R. (1980). A general psychoevolutionary theory of emotion. In: R. Plutchik & H. Kellermann (Eds.), *Emotion: Theory, Research and Experience* (pp. 3–34). New York: Academic Press.

Poincaré, G. H. (1913). *Dernières Pensées.* Paris: Flammarion.

Poland, W. (1984). The analyst's words: Empathy and countertransference. *The Psychoanalytic Quarterly, 53*: 421–424.

Pollock, G. H. (1977). Mourning: Psychoanalytic theory. In: B. Wolman (Ed.), *International Encyclopedia of Psychiatry, Psychology, Psychoanalysis and Neurology, Vol. 7* (pp. 368–371). London: Van Nostrand Reinhold.

Pongratz, L. J. (1967). *Problemgeschichte der Psychologie.* Bern/Munich: Francke.

Pontalis, J. B. (1978). Ida y vuelta. In: D. W. Winnicott, A. Green, O. Mannoni, et al., *Donald W. Winnicott.* Buenos Aires: Editorial Trieb.

Popper, K. (1963). *Conjecture and Refutation.* London: Routledge & Kegan Paul.

Pribram, K. H. (1984). Emotion: A neurobehavioral analysis. In: K. R. Scherer & P. Ekman (Eds.), *Approaches to Emotion* (pp. 13–18). Hillsdale, NJ: Lawrence Erlbaum.

Rank, O. (1914). *The Double. A Psychoanalytic Study.* London: Karnac Books, 1989.

___ (1924). *The Trauma of Birth.* New York: Brunner, 1952.

___ (1926). Technique of psychoanalysis. *Archives of Psychoanalysis, 1* (4) (July 1927).

Rapaport, D. (1938). The recent history of the association concept. *Psychoanalytical Studies, 2:* 159–180. [Reprinted in: M. M. Gill (Ed.), *The Collected Papers of David Rapaport* (pp. 37–51). New York: Basic Books, 1967.]

___ (1953). On the psychoanalytic theory of affects. *International Journal of Psycho-Analysis, 34:* 177–198. [Reprinted in: M. M. Gill (Ed.), *The Collected Papers of David Rapaport* (pp. 476–512). New York: Basic Books, 1967.]

Reich, W. (1933). *The Mass Psychology of Fascism.* New York: Orgone Institute Press, 1946.

Reik, Th. (1948). *The Inner Experience of a Psychoanalyst.* London: Allen & Unwin.

Ricoeur, P. (1965). *Freud and Philosophy. An Essay on Interpretation.* New Haven/London: Yale University Press, 1970.

Roazen, P. (1969). *Brother Animal. The Story of Freud and Tausk.* New York: A. A. Knopf.

___ (1971). *Freud and His Followers.* New York: A. A. Knopf.

___ (1985). *Helene Deutsch. A Psychoanalyst's Life.* Garden City, NY: Anchor Press/Doubleday.

Robert, P. (1989). *Le petit Robert. Dictionnaire alphabétique et analogique de la lange française.* Paris: Le Robert.

Róheim, G. (1943). *The Origin and Function of Culture.* New York: Nervous and Mental Disease Publishing, Monograph No. 69, VI.

Rorty, R. (1980). *Philosophy and the Mirror of Nature*. Oxford: Blackwell.

Rosolato, G. (1975). L'axe narcissique des dépressions. *Nouvelle Revue de Psychanalyse, 11*: 5–34.

Ross, W. D., & Kapp, F. T. (1962). A technique for self-analysis of countertransference. *Journal of the American Psychoanalytic Association, 10*: 643–657.

Rothenberg, A. (1979). *The Emerging Goddess. The Creative Process in Art, Science, and Other Fields*. Chicago/London: University of Chicago Press.

Roudinesco, E. (1982). *La bataille de cent ans. Histoire de la psychanalyse en France. Vol. 1, 1885–1939*. Paris: Ramsay.

___ (1986). *Jacques Lacan and Co. A History of Psychoanalysis in France, 1925–1985*. London: Free Association Books, 1990.

Roustang, F. (1976). *Dire Mastery. Discipleship from Freud to Lacan*. Washington, DC: American Psychiatric Press, 1982/1986.

Russell, B. (1929). *Mysticism and Logic*. New York: Norton.

Sandler, J. (1976). Countertransference and role-responsiveness. *International Review of Psycho-Analysis, 3*: 43–47.

Sapir, E. (1921). *Language*. New York: Harcourt, Brace & World.

___ (1929). *Selected Writings of Edward Sapir*. Berkeley: University of California Press.

Saussure, R. de (1956). Sigmund Freud. *Schweizerische Zeitschrift für Psychologie, 15*: 136–139. [Reprinted in: H. M. Ruitenbeek (Ed.), *Freud as We Knew Him*. Detroit: Wayne State University Press, 1973.]

Schachter, S., & Singer, J. (1962). Cognitive, social and physiological determinants of emotional states. *Psychological Revue, 63*: 379–399.

Schank, R. C. (1972). Conceptual dependency: A theory of natural language understanding. *Cognitive Psychology, 3*: 552–631.

Schank, R. C., & Abelson, R. (1977). *Scripts, Plans, Goals, and Understanding*. Hillsdale, NJ: Lawrence Erlbaum.

Scherer, K. (1982). *Vokale Kommunikation*. Weinheim: Beltz.

Schnitzler, A. (1889). Die Frage an der Schicksal. In: *Gesammelte Werke, Das dramatische Werk, Vol. 1*. Frankfurt: Fischer, 1977.

___ (1898). Paracelse. In: *Gesammelte Werke, Das dramatische Werk, Vol. 2*. Frankfurt: Fischer, 1978.

___ (1901). *Leutnant Gustel*. Berlin: Fischer.

___ (1917). *Flucht in die Finsternis*. Berlin: Fischer, 1931.

___ (1939). *Ueber Krieg und Frieden.* Posthumous Edition. Stockholm: Bermann-Fischer.

___ (1961). *Gesammelte Werke.* Frankfurt am Main: Fischer.

___ (1968). *Jugend in Wien. Eine Autobiographie.* Vienna: Fritz Molden. [Reprinted Frankfurt am Main: Fischer, 1981.]

Schnitzler, H. (1955). Sigmund Freud: Briefe an Arthur Schnitzler. *Die Neue Rundschau, 66:* 95–106.

Schofield, W. (1964). *Psychotherapy: The Purchase of Friendship.* Englewood Cliffs, NJ: Prentice Hall.

Schur, M. (1972). *Freud: Living and Dying.* New York: International Universities Press.

Searle, J. (1969). *Speech Acts.* Cambridge: Cambridge University Press.

___ (1979a). Metaphor. In: A. Ortony (Ed.), *Metaphor and Thought.* Cambridge: Cambridge University Press.

___ (1979b). *Expression and Meaning. Studies in the Theory of Speech Acts.* Cambridge/London: Cambridge University Press.

Searles, H. (1975). The patient as the therapist of his analyst. In: *Countertransference and Related Subjects* (pp. 380–459). New York: International Universities Press, 1978.

Sennett, R. (1974). *The Fall of Public Man.* New York: A. A. Knopf.

Sharpe, E. F. (1930). The technique of psycho-analysis. Seven Lectures. I. The analyst. Essential qualifications for the acquisition of technique. *International Journal of Psycho-Analysis, 11:* 251–280.

Singer, J. L. (1989). *Repression and Dissociation: Defence Mechanisms and Personality Styles.* Chicago, IL: University of Chicago Press.

Skinner, B. F. (1974). *About Behaviorism.* New York: Random House.

Slakter, E. (1987). *Countertransference.* North Vale, NJ: Jason Aronson.

Smirnoff, V. (1966). *La psychanalyse de l'enfant.* Paris: Presses Universitaires de France.

Smith, M. L., Glass, G. V., & Miller, Th. I. (1980). *The Benefits of Psychotherapy.* Baltimore/London: Johns Hopkins University Press.

Socarides, C. W. (1977). *The World of Emotions: Clinical Studies of Affects and Their Expression.* New York: International Universities Press.

Sperber, D. (1975). *Rethinking Symbolism.* Cambridge: Cambridge University Press.

Sperber, D., & Wilson, D. (1986): *Relevance. Communication and Cognition.* Oxford: Blackwell.

Spielrein, S. (1911). Ueber den psychologischen Inhalt eines Falles von Schizophrenia. *Jahrbuch für psychopathologische und psychoanalytische Forschungen, 3:* 329–400.

___ (1912). La destruction comme cause du devenir. In: A. Carotenuto & C. Trombetta, *Sabina Spielrein entre Freud et Jung* (pp. 213–262). Paris: Aubier Montaigne, 1981.

___ (1980). Diary. In: A. Carotenuto, *A Secret Symmetry. Sabina Spielrein Between Freud and Jung* (pp. 3–44). London: Routledge & Kegan Paul, 1984.

___ (1987). *Sämtliche Schriften.* Freiburg i. Br.: Kore.

Squire, L. R. (1987). *Memory and Brain.* Oxford: Oxford University Press.

Steele, R. S. (1982). *Freud and Jung. Conflicts of Interpretation.* London/Boston: Routledge & Kegan Paul.

Steiner, R. (1985). Some thoughts about tradition and change arising from an examination of the British Psychoanalytical Society's Controversial Discussions (1943–1944). *International Review of Psycho-Analysis, 12* (1): 27–71.

Stekel, W. (1924). The polyphony of thought. In: D. Rapaport, *Organization and Pathology of Thought.* New York: Columbia University Press, 1951.

Sterba, R. F. (1934). The fate of the ego in analytic therapy. *International Journal of Psycho-Analysis, 15:* 117–126.

Stern, D. (1985). *The Interpersonal World of the Infant. A View from Psychoanalysis and Developmental Psychology.* New York: Basic Books.

Stewart, H. (1987). The varieties of transference interpretation. *International Journal of Psycho-Analysis, 68:* 279–297.

Stoller, R. J. (1975). *Perversion. The Erotic Form of Hatred.* New York: Pantheon Books, Random House. [Reprinted London: Karnac Books, 1986.]

___ (1979). *Sexual Excitement. Dynamics of Erotic Life.* New York: Pantheon Books, Random House. [Reprinted London: Karnac Books, 1986.]

Strachey, J. (1934). The nature of the therapeutic action of psycho-analysis. *International Journal of Psycho-Analysis, 15:* 127–159.

Strupp, H. H., Wallach, M. S., Wogan, M., & Jenkins, J. W. (1963). Psychotherapists' assessments of former patients. *Journal of Nervous and Mental Disease, 137:* 222–230.

Suyka, O. (1905). Zwei Bücher. *Die Fackel, 191* (21 December): 6–11.

Szasz, T. S. (1963). The concept of transference. *International Journal of Psycho-Analysis, 44*: 432–443.

___ (1977). *Karl Kraus and the Soul-Doctors.* London: Routledge & Kegan Paul.

Thom, R. (1974). *Modèle mathématique de la morphogenèse.* Série 10/18.

Thomas, D. M. (1981). *The White Hotel.* New York: Viking Press.

Tomkins, S. S. (1984). Affect theory. In: K. R. Scherer & P. Ekman (Eds.), *Approaches to Emotion* (pp. 163–196). Hillsdale, NJ: Lawrence Erlbaum.

Tyler, S. A. (1969). *Cognitive Anthropology.* New York: Rinehart & Winston.

Uexküll, Th. von (Ed.) (1986). *Psychosomatische Medizin* (third edition). Munich: Urban & Schwartzenberg.

Viderman, S. (1970). *La construction de l'espace analytique.* Paris: Denoël.

Waelder, R. (1930). The principle of multiple function. Observations on over-determination. *The Psychoanalytic Quarterly, 5* (1936): 45–62.

Wallerstein, R. S. (1986). *Forty-Two Lives in Treatment. A Study of Psychoanalysis and Psychotherapy.* New York/London: The Guilford Press.

Wedekind, F. (1891). *Frühlingserwachen.* Zurich: Jean Gross.

___ (1895). *Erdgeist.* Munich: A. Langen.

___ (1904). *Die Büchse der Pandora.* Berlin: Cassirer.

___ (1905). *Tod und Teufel (Totentanz). Drei Szenen.* Reprint: *Totentanz. Drei Szenen.* Munich: A. Langen, 1906.

___ (1924). *Gesammelte Werke.* Munich: G. Müller.

Weinberger, D. (1989). The construct validity of the repressive coping style. In: J. L. Singer (Ed.), *Repression and Dissociation: Defensive Mechanisms and Personality Styles.* Chicago, IL: University of Chicago Press.

Whybrow, P. (1984). Contributions from neuroendocrinology. In: K. R. Scherer & P. Ekman (Eds.), *Approaches to Emotion* (pp. 59–72). Hillsdale, NJ: Lawrence Erlbaum.

Whyte, L. L. (1960). *The Unconscious Before Freud.* New York: Basic Books.

Wilson, C. (1988). *Rudolf Steiner, visionnaire au coeur de l'homme.* Paris: Le Rocher.

Winnicott, D. W. (1949). Hate in the countertransference. *International Journal of Psycho-Analysis, 30*: 69–74. [Also in *Through Paediatrics to Psychoanalysis*. London: Tavistock, 1958. Reprinted London: Karnac Books, 1992.]

—— (1956). On transference. *International Journal of Psycho-Analysis, 37*: 386–388. [Republished as: Clinical varieties of transference. In: *Through Paediatrics to Psychoanalysis*. London: Tavistock, 1958. Reprinted London: Karnac Books, 1992.]

Wittels, F. (1910). *Ezechiel der Zugereiste*. Berlin: Egon Fleischel.

Wittgenstein, L. (1931). *Culture and Value*. Chicago, IL: University of Chicago Press, 1980.

Wohl, R. (1979). *The Generation of 1914*. New York: Harvard University Press.

Wollheim, R. (1979). The cabinet of Lacan. *The New York Review* (26 January).

Wurmser, L. (1987). *Flucht vor dem Gewissen. Analyse von Ueber-Ich und Abwehr bei schweren Neurosen*. Berlin: Springer Verlag.

Zimmer, D. E. (1986). *Tiefenschwindel: Die endlose und die beendbare Psychoanalyse*. Reinbek bei Hamburg: Rowohlt.

Zweig, S. (1947). *World of Yesterday. An Autobiography*. London: Cassell, 1987.

Zwiebel, R. (1985). The dynamics of the countertransference dream. *International Review of Psycho-Analysis, 12* (1): 87–99.